Mastering LangChain

A Comprehensive Guide to Building Generative AI Applications

Sanath Raj B Narayan
Nitin Agarwal

Apress®

Mastering LangChain: A Comprehensive Guide to Building Generative AI Applications

Sanath Raj B Narayan
Bangalore, Karnataka, India

Nitin Agarwal
Bangalore, Karnataka, India

ISBN-13 (pbk): 979-8-8688-1717-5
https://doi.org/10.1007/979-8-8688-1718-2

ISBN-13 (electronic): 979-8-8688-1718-2

Copyright © 2025 by Sanath Raj B Narayan and Nitin Agarwal

This work is subject to copyright. All rights are reserved by the Publisher, whether the whole or part of the material is concerned, specifically the rights of translation, reprinting, reuse of illustrations, recitation, broadcasting, reproduction on microfilms or in any other physical way, and transmission or information storage and retrieval, electronic adaptation, computer software, or by similar or dissimilar methodology now known or hereafter developed.

Trademarked names, logos, and images may appear in this book. Rather than use a trademark symbol with every occurrence of a trademarked name, logo, or image we use the names, logos, and images only in an editorial fashion and to the benefit of the trademark owner, with no intention of infringement of the trademark.

The use in this publication of trade names, trademarks, service marks, and similar terms, even if they are not identified as such, is not to be taken as an expression of opinion as to whether or not they are subject to proprietary rights.

While the advice and information in this book are believed to be true and accurate at the date of publication, neither the authors nor the editors nor the publisher can accept any legal responsibility for any errors or omissions that may be made. The publisher makes no warranty, express or implied, with respect to the material contained herein.

Managing Director, Apress Media LLC: Welmoed Spahr
Acquisitions Editor: Celestin Suresh John
Development Editor: Laura Berendson
Editorial Assistant: Gryffin Winkler

Cover designed by eStudioCalamar

Cover image designed by Freepik (www.freepik.com)

Distributed to the book trade worldwide by Springer Science+Business Media New York, 1 New York Plaza, New York, NY 10004. Phone 1-800-SPRINGER, fax (201) 348-4505, e-mail orders-ny@springer-sbm.com, or visit www.springeronline.com. Apress Media, LLC is a Delaware LLC and the sole member (owner) is Springer Science + Business Media Finance Inc (SSBM Finance Inc). SSBM Finance Inc is a **Delaware** corporation.

For information on translations, please e-mail booktranslations@springernature.com; for reprint, paperback, or audio rights, please e-mail bookpermissions@springernature.com.

Apress titles may be purchased in bulk for academic, corporate, or promotional use. eBook versions and licenses are also available for most titles. For more information, reference our Print and eBook Bulk Sales web page at http://www.apress.com/bulk-sales.

Any source code or other supplementary material referenced by the author in this book is available to readers on GitHub. For more detailed information, please visit https://www.apress.com/gp/services/source-code.

If disposing of this product, please recycle the paper

Table of Contents

About the Authors .... **xi**

About the Technical Reviewer .... **xiii**

Chapter 1: Introduction to LangChain **1**

    What Is LangChain? 1

    Evolution of Language Models 2

    Key Features and Capabilities of LangChain 2

    The Role of LangChain 4

    Real-World Use Cases 4

    A Quick Start Guide to LangChain 5

        Setting Up Your Environment 5

        Installation and Setup 5

        Building Your First Chain 6

    Exploring the LangChain Ecosystem 7

    Key Takeaways 9

Chapter 2: Core Components of LangChain **11**

    Chains 11

        Key Components of Chains 12

        Why Do We Need Chains? 12

        Types of Chains 14

        Designing Effective Chains 32

    Prompt Templates 37

        Understanding the Importance of Prompts in LLMs 37

        Creating and Customizing Prompt Templates 38

        Dynamic Prompt Generation 39

        Prompt Optimization Techniques 40

TABLE OF CONTENTS

Managing Prompt Libraries	42
Tools and Function Calling	43
Overview of Tools in LangChain	43
Built-in Tools and Their Functionalities	44
Creating Custom Tools	46
Integrating External APIs As Tools	49
Function Calling: Enhancing LLM Capabilities	53
Key Takeaways	60

Chapter 3: Advanced Components and Integrations 61

Output Parser	62
Choosing the Right Parser	64
Structured Output Parsing	64
Custom Output Parsers for Specific Data Formats	67
Error Handling in Output Parsing	69
Integrating Parsers with Chains and Models	71
Memory Components	73
Understanding the Role of Memory in LangChain	73
Types of Memory in LangChain	73
Implementing Memory in Chains and Agents	74
Managing Long-Term Memory and Context	77
Embeddings and Vector Stores	80
Embeddings in LangChain	81
Types of Embedding Models Supported	81
Creating and Managing Vector Stores	84
Semantic Search and Similarity Matching	89
Agents	90
Types of Agents	91
Agent Execution and Decision-Making Process in LangChain	98
Customization Options and Extending Agent Capabilities	100
Extending Capabilities	100
Callbacks and Logging	102

Chat Models and LLMs 102
    Differences Between Chat Models and LLMs 102
    Supported Chat Models in LangChain 103
    Configuring and Fine-Tuning Chat Models 105
LangChain Expression Language (LCEL) 107
    Example 1: Basic LCEL Syntax 107
    Eample 2: LCEL Allows for the Creation of More Sophisticated Chains 108
    Eample 3: Using RunnableParallel for Multiple Inputs LCEL Supports Parallel Operations . 108
    Example 4: Error Handling in LCEL 109
    Example 5: Using LCEL with Retriever 110
    Example 6: LCEL Advanced Chain 111
Key Takeaways 113

Chapter 4: Building Chatbots 115
Why Use LangChain for Chatbots? 115
    Key Advantages 115
Understanding Conversation Flows 119
    Components of a Conversation Flow 119
Building a Simple Chatbot with LangChain 120
    Step-by-Step Guide to Build a Simple Chatbot 121
Implementing Context Awareness in Conversations 125
    Why Context Awareness Matters 125
    Best Practices for Context-Aware Chatbots 126
Handling Complex Queries and Multi-turn Dialogues 126
    Complex Queries 126
    Multi-turn Dialogues 128
    Best Practices 130
Key Takeaways 130

Chapter 5: Building Retrieval- Augmented Generation (RAG) Systems 131
Overview of the RAG 131
    Components of a RAG System 132
    Approach to RAG Implementation 137

TABLE OF CONTENTS

Use Cases and Applications of RAG 138
Data Loading and Preprocessing 140
Techniques for Efficient Data Loading 141
Data Cleaning and Normalization 141
Handling Different Types of Data (Text, PDFs, Web Content) 142
Chunking Strategies 144
Fixed-Length Chunking 144
Semantic Chunking Techniques 145
Sentence- and Paragraph-Based Chunking 146
Overlapping Chunks and Sliding Windows 147
Optimizing Chunk Size for Different Use Cases 147
Embedding Data 149
Introduction to Text Embeddings 149
Embedding Models Supported in LangChain 149
Generating Embeddings for Chunks 152
Handling Large-Scale Embedding Tasks 152
Indexing: Vector Stores 154
Types of Vector Stores Supported in LangChain 154
Creating and Managing Vector Indices 155
Scalability and Performance Considerations 155
Choosing the Right Vector Store for Your Application 155
Retrieval Techniques 156
Similarity Search Algorithms 156
Dense Retrieval vs. Sparse Retrieval 157
Hybrid Retrieval Approaches 158
Implementing Custom Retrieval Methods 158
Metadata Filtering and Faceted Search 159
Improving Model Retrieval 159
Query Expansion and Reformulation 159
Re-ranking Retrieved Documents 160
Relevance Feedback Mechanisms 161

TABLE OF CONTENTS

    Fine-Tuning Retrieval Models 161

    Ensemble Methods for Improved Retrieval 162

Response Generation Using LLMs 163

    Integrating Retrieved Context with LLM Prompts 164

    Prompt Engineering for RAG Systems 164

    Handling Multi-turn Conversations in RAG 165

    Balancing Retrieved Information and Model Knowledge 166

    Techniques for Maintaining Coherence and Relevance 167

Ethical Considerations and Best Practices 168

    Handling Sensitive Information in RAG Systems 168

    Bias Mitigation in Retrieval and Generation 169

    Transparency and Explainability in RAG 170

    Data Privacy and Compliance Considerations 170

    Conclusion 171

Key Takeaways 171

Chapter 6: LangServe, LangSmith, and LangGraph: Deploying, Optimizing, and Designing Language Model Workflows 173

LangServe 175

    Uses of LangServe 175

LangGraph 178

    Uses of LangGraph 178

LangSmith 180

    Application of LangSmith with LangChain Deployments 180

    Streamlining AI Development with LangSmith 181

    Setting Up and Managing Projects with LangSmith 181

Key Takeaways 182

Chapter 7: LangChain and NLP 183

NLP Techniques in LangChain 183

    Sentiment Analysis and Classification 183

    Fine-Tuning for Specific Tasks 191

Key Takeaways 198

Chapter 8: Building AI Agents with LangGraph 199
- Core Components of LangGraph 200
  - Graphs 200
  - States and State Management 200
  - Nodes 201
  - Edges and Conditional Routing 201
- Basic Structure of a LangGraph Agent 202
- Creating Autonomous Agents Using LangGraph 203
  - Defining Agent State for Autonomous Agents 204
- Agent Architectures 206
  - ReAct Architecture 206
  - Reflexion Architecture 206
  - Plan-and-Execute Architecture 207
  - Multi-agent Architectures 207
- Case Study 207
  - Reflection-Based Agents for Content Generation 207
  - Agentic Text-to-SQL Generator Using LangGraph 208
- Key Takeaways 209

Chapter 9: LangChain Framework Integration 211
- Working with External APIs 211
  - API Integration Best Practices 211
  - Using LangChain with Popular APIs 212
- Combining LangChain with Other Frameworks 214
  - TensorFlow, PyTorch, and More 214
  - Hybrid AI Architectures 216
- Key Takeaways 220

Chapter 10: Deploying LangChain Applications 221
- Preparing for Production 221
- Architecture 222
- Scaling Applications 223

Optimizing for Production	224
Testing and Evaluating Applications	225
Monitoring and Logging	226
Key Takeaways	227

Chapter 11: Best Practices and Practical Aspects 229

Building Ethical and Compliant AI Systems	229
Ethical Considerations in AI Development	229
Navigating Regulatory Compliance	230
Optimizing and Scaling LangChain Applications	231
Performance Tuning Strategies	231
Scaling Solutions for Production Readiness	231
Avoiding Common Pitfalls	232
Typical Mistakes in LangChain Projects	232
Proactive Strategies for Risk Mitigation	233
The Future of LangChain and AI Agents	233
Emerging Technologies Shaping LangChain	233
LangChain Roadmap and Community Vision	234
Preparing for the Next Generation of AI Development	234
Embracing Continuous Learning	234
Staying Competitive in a Rapidly Evolving Landscape	235
Key Takeaways	236

Index 237

About the Authors

Sanath Raj Narayan is a Senior Data Scientist with over a decade of experience in building AI and machine learning solutions, as well as scalable systems using AWS and Azure. He has previously held roles at Ericsson, Mindtree, KPMG India, and Cognizant, where he led data-driven projects across the retail, telecom, and consulting sectors. Sanath Raj's expertise spans predictive modeling, recommender systems, and the deployment of end-to-end machine learning pipelines. He is also a regular speaker at conferences, where he presents on AI and related topics.

Nitin Agarwal is a Principal AI Scientist with over 14 years of experience in Artificial Intelligence and Data Science. Formerly a Senior Data Scientist at Microsoft, he specializes in Machine Learning, Deep Learning, Natural Language Processing, and Statistical Modeling. Nitin brings extensive expertise in crafting innovative AI Copilots and delivering cutting-edge Data Science solutions across diverse industries, including Healthcare, Technology, and Logistics. He holds a master's degree in Data Science and Engineering from Birla Institute of Technology and Sciences (BITS), Pilani, and CORe from Harvard Business School (HBX). Passionate about Generative AI and Large Language Models (LLMs), he is also a published researcher and a dedicated mentor. Nitin frequently shares his expertise as a speaker at AI and technology conferences, where he engages with the community on the latest advancements in AI and their real-world applications.

About the Technical Reviewer

Rutvik Acharya is a Machine Learning and AI professional with over 13 years of experience in the technology industry. Currently serving as a Principal at Atlassian, he leads multiple end-to-end AI projects and drives the company's artificial intelligence strategy across the organization. His expertise spans across various domains of artificial intelligence, including natural language processing, machine learning operations, and enterprise AI solutions. Throughout his career, he has been instrumental in developing and implementing AI strategies that bridge the gap between cutting-edge technology and practical business applications. As an active participant in the AI community, he regularly contributes to discussions about the future of artificial intelligence and its impact on business transformation. His blend of technical expertise and strategic insight provides him with a unique perspective on the evolving landscape of AI and its real-world applications.

CHAPTER 1

Introduction to LangChain

In this chapter, you'll embark on a foundational journey into the world of LangChain, a powerful framework designed to streamline the development of applications powered by large language models (LLMs). We begin by demystifying what LangChain is and how it fits into the rapidly evolving landscape of language models. You'll explore LangChain's core features, understand its purpose, and examine real-world use cases that highlight its practical value. The chapter also walks you through a quick start guide to help you get up and running, followed by an overview of the broader LangChain ecosystem. Whether you're a beginner or transitioning from traditional NLP workflows, this chapter sets the stage for building intelligent, context-aware LLM applications with LangChain.

What Is LangChain?

LangChain represents a groundbreaking framework that has fundamentally transformed the landscape of artificial intelligence development, particularly in the domain of language model applications. At its core, LangChain is an open source Python library designed to simplify and enhance the process of building sophisticated applications powered by large language models (LLMs). Unlike traditional programming frameworks, LangChain provides a unique abstraction layer that allows developers to create complex, context-aware AI applications with unprecedented ease and flexibility.

The fundamental philosophy behind LangChain is to address the inherent limitations of standalone language models by introducing a modular and composable approach to AI application development. It acts as a comprehensive toolkit that enables developers to chain together multiple components, creating intelligent systems that can perform complex tasks beyond the capabilities of individual language models.

CHAPTER 1   INTRODUCTION TO LANGCHAIN

Evolution of Language Models

To fully appreciate LangChain's significance, we must first understand the evolutionary journey of language models. The field of natural language processing (NLP) has undergone a dramatic transformation over the past decade. Early language models were relatively simplistic, relying on statistical approaches and limited context understanding. These models could perform basic tasks like text completion and simple translations but lacked the nuanced comprehension and generative capabilities we see today.

The breakthrough came with the introduction of transformer architectures and large language models like the GPT (Generative Pre-trained Transformer) series, BERT, and their successors. These models demonstrated an unprecedented ability to understand and generate human-like text, drawing from massive training datasets. However, they also revealed significant challenges:

- Limited contextual memory
- Difficulty in performing complex, multi-step tasks
- Lack of external knowledge integration
- Challenges in maintaining consistent reasoning
- Hallucinations and inaccuracy
- Outdated training data

LangChain emerged as a direct response to these limitations, providing a sophisticated framework that addresses these fundamental challenges in language model applications.

Key Features and Capabilities of LangChain

LangChain offers a multitude of benefits that set it apart from traditional AI development approaches:

1. **Modular Component Architecture**
   - Enables developers to create flexible, reusable AI components
   - Supports easy combination and reconfiguration of different AI building blocks
   - Allows seamless integration of various language models, memory systems, and external tools

2. **Advanced Prompt Management**
   - Provides sophisticated prompt templates and management systems
   - Supports dynamic prompt generation based on context
   - Enables complex prompt chaining and transformation

3. **Comprehensive Memory Systems**
   - Implements various memory types including conversational, entity, and knowledge-based memory
   - Allows persistent state management across different AI interactions
   - Supports both short-term and long-term memory configurations

4. **Seamless External Tool Integration**
   - Enables easy connection with external APIs, databases, and computational tools
   - Supports real-time data retrieval and context augmentation
   - Facilitates complex reasoning across multiple information sources

5. **Retrieval Augmented Generation (RAG)**
   - Supports integration with vector stores and retrievers for grounded, context-aware responses
   - Enhances factual accuracy and reduces hallucinations by incorporating external knowledge

6. **Agentic Functionality**
    - Enables multi-step, goal-driven task execution through autonomous and human-in-the-loop agents
    - Supports dynamic orchestration of tools, memory, and reasoning steps across agent workflows

The Role of LangChain

LangChain has quickly become a pivotal framework in the AI development ecosystem. It bridges the gap between powerful language models and practical, real-world applications. By providing a flexible and extensible architecture, LangChain empowers developers to create AI solutions that go far beyond simple text generation.

The framework's significance lies in its ability to transform monolithic language models into intelligent, context-aware systems capable of

- Performing complex, multi-step reasoning
- Integrating external knowledge sources
- Maintaining contextual awareness
- Adapting to diverse application requirements

Real-World Use Cases

To illustrate LangChain's versatility, consider these compelling use cases:

1. **Intelligent Customer Support Systems**
    - Develop AI agents that can understand complex customer queries
    - Retrieve relevant information from knowledge bases
    - Provide personalized, contextually accurate responses

2. **Research and Analysis Tools**
   - Create AI-powered research assistants
   - Synthesize information from multiple sources
   - Generate comprehensive reports and summaries

3. **Code Generation and Debugging Assistants**
   - Build intelligent coding companions
   - Provide context-aware code suggestions
   - Assist in debugging and documentation

A Quick Start Guide to LangChain

Setting Up Your Environment

Before diving into LangChain development, ensure you have the following:

1. Python 3.11 or higher
2. Basic understanding of Python programming
3. Familiarity with pip package manager
4. OpenAI API key (optional, but recommended for advanced features)
5. Jupyter Notebook or preferred Python IDE

Installation and Setup

Installing LangChain is straightforward. Use pip to install the core library:

```
pip install langchain
```

How to get the API key

- **Visit the API Keys page**: https://platform.openai.com/api-keys.
- **Click "+ Create new secret key"** and give it a name.

For additional dependencies, you might want to install specific integrations:

```
# OpenAI integration
pip install langchain-openai
# Community extensions
pip install langchain-community
```

Building Your First Chain

Here's a simple example to demonstrate LangChain's power:

```
from langchain.chains import LLMChain
from langchain.prompts import PromptTemplate
from langchain_openai import OpenAI

# Initialize the language model
llm = OpenAI(temperature=0.7)

# Create a prompt template
prompt = PromptTemplate(
    input_variables=["topic"],
    template="Write a brief, engaging paragraph about {topic}"
)

# Create the chain
chain = llm | prompt

# Run the chain
result = chain.invoke(["artificial intelligence"])
print(result)
```

Output:
{'topic': ['artificial intelligence'], 'text': '\n\nArtificial intelligence, also known as AI, is a rapidly growing field that is revolutionizing the way we live, work, and interact with technology. It is the development of computer systems that can perform tasks that normally require human intelligence, such as problem-solving and decision-making. AI is already being used in various industries, from self-driving cars to personalized recommendations on streaming services. While some may fear the

rise of AI, it has the potential to greatly enhance our lives and improve efficiency in many areas. The possibilities for artificial intelligence are endless, and the future is full of exciting developments in this cutting-edge field.'}

The "chain.invoke()" method, when used with "LLMChain" and "PromptTemplate" like this, returns a dictionary where the key is the output variable name from the prompt template (in this case, "text"). The output shown includes the input variable "topic" in the dictionary as well, which is not the standard output structure for this basic chain.

Exploring the LangChain Ecosystem

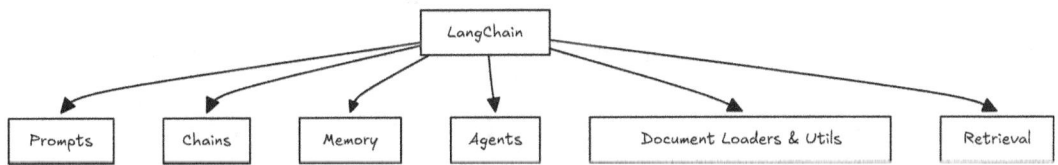

Figure 1-1. LangChain ecosystem

Figure 1-1 covers major components of the LangChain ecosystem. LangChain is a comprehensive framework that enables developers to build sophisticated AI applications by combining various components into cohesive solutions. Its architecture is designed to be modular, flexible, and powerful enough to handle complex AI tasks while remaining accessible to developers.

Key Components

1. **Prompts and Templates:** At the foundation of LangChain are prompts and templates, which provide structured ways to interact with language models. These components help developers create consistent and effective interactions by

   - Defining reusable prompt patterns for common tasks
   - Implementing dynamic templates that adapt to specific contexts
   - Managing prompt variations and optimization strategies

2. **Chains:** These represent the workflow engine of LangChain, allowing developers to combine multiple components into logical sequences. They provide

    - Structured patterns for connecting different components
    - Sequential processing of complex tasks
    - Integration points between prompts, models, and outputs

3. **Memory:** The memory system ensures continuity and context retention across interactions. It includes

    - Vector database integration for efficient storage
    - Context management across conversations
    - Flexible storage solutions for different types of information

4. **Agents:** They serve as autonomous decision-makers within the LangChain ecosystem. They can

    - Execute complex tasks independently
    - Use various tools and resources to solve problems
    - Make decisions based on context and instructions

5. **Document Loaders and Utilities:** These components handle the practical aspects of working with different data sources:

    - Supporting multiple file formats and data types
    - Providing preprocessing and transformation tools
    - Managing data pipeline workflows

6. **Retrieval:** The retrieval system enables efficient access to information and knowledge:

    - Implementation of RAG (Retrieval-Augmented Generation) systems
    - Advanced search capabilities
    - Integration with external knowledge bases

Each component is designed to work seamlessly with others, allowing developers to create sophisticated applications that can handle everything from simple queries to complex, multi-step processes. The modular nature of LangChain means that developers can start with basic implementations and gradually add complexity as their needs evolve.

This ecosystem continues to grow and adapt to new developments in AI technology, making it a powerful framework for building the next generation of AI applications.

Key Takeaways

- LangChain is a specialized framework for building end-to-end LLM-powered applications.

- It sits at the intersection of modern LLMs and application development by offering structured components like chains, tools, and memory.

- The framework provides a modular and extensible approach to integrate LLMs with external data, APIs, and tools.

- Real-world use cases show versatility across domains like customer support, code generation, and RAG systems.

- A quick start guide and ecosystem overview equip readers with the knowledge to begin experimenting with LangChain immediately.

In the next chapter, we'll take a deeper dive into the core building blocks of LangChain, starting with Chains, Prompt Templates, and Tool integrations, to understand how to construct powerful and composable LLM workflows.

CHAPTER 2

Core Components of LangChain

LangChain introduces a robust abstraction over large language models (LLMs), transforming isolated predictions into contextual, purpose-driven applications. In this chapter, we explore the heart of LangChain—Chains, Prompt Templates, and Tools and Function Calling—and understand how they empower developers to build intelligent and modular workflows.

Chains

Like the name suggests, LangChain is the framework that acts as a chain and binds across various components around the large language models (LLMs).

Chains refer to sequences of calls—whether to an LLM, a tool, or a data preprocessing step. It is an end-to-end wrapper around multiple individual components executed in a defined order. Chains are one of the core concepts of LangChain. Chains allow you to go beyond just a single API call to a language model and instead chain together multiple calls in a logical sequence. They allow you to combine multiple components to create a coherent application.

The main purpose of using Chain as compared to just using the LLM is, it helps in combining or chaining multiple prompts together, while directly passing a prompt only allows one. With Chains we also can break down a complex prompt into multiple more prompts. We can also maintain state and memory between prompts. The output of one step can be given as input to the other component; this can give additional context while directly passing prompts lack this memory. Also adding preprocessing logic, validation between prompts is easier. This keeps the developer in control of the output from the model and quality control.

CHAPTER 2 CORE COMPONENTS OF LANGCHAIN

Key Components of Chains

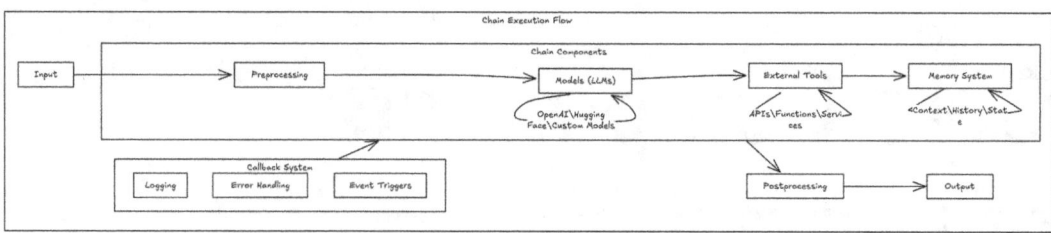

Figure 2-1. *Key components of chains*

We can see in Figure 2-1 that chains can include various components like

- **Models:** Language models (e.g., OpenAI, Hugging Face) that generate text.

- **Tools:** External APIs or functions that perform specific tasks (e.g., fetching data, processing images).

- **Memory:** Stores information across interactions, enabling the system to remember context or previous inputs.

- **Inputs and Outputs:** Each component in the chain takes an input, processes it, and passes the output to the next component. This flow can involve text, structured data, or any other type of information.

- **Callbacks:** Chains can utilize callbacks for tasks like logging, error handling, or triggering additional actions based on specific conditions.

Why Do We Need Chains?

Chains allow for a modular approach, where different functionalities (like data processing, model inference, and API calls) can be encapsulated in individual components. Chains allow for quick iteration and prototyping; this allows developers to easily add or remove components to test different approaches or algorithms and also by chaining components.

CHAPTER 2   CORE COMPONENTS OF LANGCHAIN

We can break down complex logic into manageable steps, which helps in easier identification of issues by isolating steps in the chain and also simplifying updates and modifications to individual components without overhauling the entire system. Chains can easily integrate various external tools, APIs, and services. While building complex applications, managing state and context is essential. Chains can incorporate memory components that allow the system to remember past interactions.

Also, chains enable dynamic and flexible workflows which gives us the options for

- **Sequential Processing:** Pass the output of one component directly to the next, making it easy to create linear workflows.

- **Conditional Logic:** Implement branching logic, where the next step depends on the results of previous steps (e.g., different processing based on user input).

- **Parallel Execution:** Execute multiple components simultaneously, which can improve efficiency for tasks that can run concurrently.

Initializing Azure OpenAI

To work with Azure OpenAI in LangChain, you'll need to initialize the AzureOpenAI model wrapper. Here's a sample code snippet:

```
from langchain_azure_openai import AzureOpenAI  # Ensure this package is installed

llm = AzureOpenAI(
    deployment_name="<Azure deployment name>",  # Replace with your
                                                actual Azure OpenAI
                                                deployment name
    model_name="<Model name>",                  # Replace with the model
                                                name (e.g., "gpt-35-turbo")
    api_key="your-azure-api-key",
    api_version="2023-05-15",
    azure_endpoint="https://<your-resource-name>.openai.azure.com/"
)
```

> **Note** Be sure to replace the placeholders <Azure deployment name> and <Model name> with your actual Azure OpenAI deployment configuration. This is crucial for the model to function correctly.

Additionally, ensure that you have the appropriate package installed:

- Use `langchain-azure-openai` if you're working with Azure OpenAI.
- Use `langchain-openai` if you're using the standard OpenAI API.

Types of Chains

Simple Chain: Basic Sequential Processing

Simple chain allows for straightforward, sequential processing of inputs. Each step in the chain takes the output from the previous step as its input.

Example Code implementation

```
from langchain import PromptTemplate
from langchain.chains import LLMChain
template = '''You are an experience cook.
list the ingredients to cook the following fooditem "{food}" in {language}.'''
prompt_template = PromptTemplate.from_template(template=template)
# Initialize the language model
llm = AzureOpenAI(deployment_name="<Azure deployment name>", model_name="<Model name>")
chain = LLMChain(
    llm=llm,
    prompt=prompt_template,
    verbose=False
)
output = chain.invoke({'food': 'pizza', 'language': 'English'})
print("output:\n", output)
```

output:
{'food': 'pizza', 'language': 'English', 'text': '\n\n- Pizza dough\n- Pizza sauce\n- Shredded mozzarella cheese\n- Toppings of your choice (e.g. pepperoni, mushrooms, bell peppers) \n- Olive oil\n- Garlic powder\n- Italian seasoning\n- Salt\n- Black pepper'}

This code uses LangChain to create a language model chain that generates ingredient lists for a specified food item in a chosen language. It defines a prompt template with placeholders for food and language, creates a prompt using PromptTemplate, initializes an Azure OpenAI language model, and sets up an LLMChain. When invoked with "pizza" and "English", it will generate a list of pizza ingredients in English.

LLM Chain: Integrating Language Models

LLMChain is specifically designed to integrate with language models. It facilitates generating text based on prompts and processing model outputs effectively.

Use Cases

- **Chatbots:** Building conversational agents that generate responses based on user queries
- **Content Creation:** Automating the generation of articles, summaries, or reports using language models
- **Question Answering:** Implementing systems that respond to user questions by leveraging a language model's capabilities

Example Code implementation sample code: A chain that takes user input as a prompt, sends it to a language model, and retrieves the generated response.

```
from langchain.chains import LLMChain
from langchain.llms import OpenAI

# Initialize the language model
model = OpenAI(modelModel name"")

# Create an LLMChain
llm_chain = LLMChain(model=model)
```

```
# Define a prompt
prompt = "What are the benefits of using LangChain for AI applications?"

# Execute the chain
result = llm_chain.run(prompt)
print(result)  # Output: (Generated response from the language model)
```

This code demonstrates using LangChain to interact with an OpenAI language model by creating an LLMChain with a specified prompt and generating a response about the benefits of LangChain for AI applications. It initializes the model, sets up the chain, runs the prompt, and prints the generated result.

Sequential Chain: Combining Multiple Chains

Sequential chain allows you to combine multiple chains together in a sequence. Each step in the chain takes the output from the previous step and uses it as input for the next one. This is particularly useful when you want to process data through a series of transformations, like summarizing a document, extracting key information, and then generating insights based on that extracted data.

Example Code Implementation

```
from langchain.chains import SequentialChain, LLMChain
from langchain.prompts import ChatPromptTemplate
from langchain_openai import AzureOpenAI

# Initialize the LLM (Azure OpenAI in this case)
llm = AzureOpenAI(deployment_name="<Azure deployment name>", model_name="<Model name>")

# Step 1: Summarize the customer feedback
template1 = "Summarize the following customer feedback:\n{feedback}"
prompt1 = ChatPromptTemplate.from_template(template1)
chain_1 = LLMChain(llm=llm, prompt=prompt1, output_key="feedback_summary")

# Step 2: Identify key issues from the feedback summary
template2 = "Identify the key issues or complaints from this feedback summary:\n{feedback_summary}"
prompt2 = ChatPromptTemplate.from_template(template2)
chain_2 = LLMChain(llm=llm, prompt=prompt2, output_key="key_issues")
```

```python
# Step 3: Generate improvement suggestions based on the key issues
template3 = "Based on these key issues, suggest improvements to the product:\n{key_issues}"
prompt3 = ChatPromptTemplate.from_template(template3)
chain_3 = LLMChain(llm=llm, prompt=prompt3, output_key="improvement_suggestions")

# Create the Sequential Chain
seq_chain = SequentialChain(chains=[chain_1, chain_2, chain_3], input_variables=['feedback'], output_variables=['feedback_summary', 'key_issues', 'improvement_suggestions'], verbose=True)

# Example Customer Feedback
customer_feedback = '''
I have been using this smartphone for about 6 months now, and while it performs well in terms of speed,
I am facing significant issues with battery life. The battery drains very quickly, especially when using apps with high performance requirements.
Also, the camera quality does not match what was advertised. I expected better photo quality, particularly in low-light conditions.
On the positive side, I appreciate the build quality and the design, but the battery and camera need major improvements.
'''

# Run the Sequential Chain
 results = seq_chain.invoke({'feedback': customer_feedback})

# Print the final improvement suggestions
print(results['improvement_suggestions'])

1. Manufacturers should focus on optimizing battery life and improving energy efficiency to meet customer demands.
2. Software integration should be enhanced for better performance and user experience.
3. Further exploration of foldable screens could provide competitive differentiation as the technology gains mainstream acceptance.
4. Emphasizing sustainability and environmentally friendly practices can attract eco-conscious customers and improve brand reputation.
```

This code demonstrates a sequential processing of customer feedback using LangChain and Azure OpenAI. It creates a three-step chain.

Explanation of Each Chain

> **Chain 1 (Summarize the Text):** Takes the input document and generates a summary of it. The output is stored as a summary.
>
> **Chain 2 (Identify Key Topics):** Takes the summary generated by Chain 1 and identifies the most important topics or themes discussed in the document. The output is stored as key_topics.
>
> **Chain 3 (Generate Actionable Insights):** Takes the key_topics from Chain 2 and generates actionable insights or suggestions based on those topics. The output is stored as insights.

Sequential chain also allows us to combine multiple LLM chains in a stepwise manner, with each chain building on the results of the previous one. This creates a flexible and powerful way to handle complex workflows.

RouterChain: Dynamic Routing Based on Input

RouterChain is a powerful component in LangChain that enables dynamic routing of inputs to different chains or tools based on the content or characteristics of the input. This chain type is particularly useful when dealing with diverse inputs that require different processing pipelines.

Key Features of RouterChain

- **Input Analysis:** Examines the input to determine the most appropriate route
- **Flexible Routing:** Can direct inputs to various chains, tools, or agents
- **Customizable Logic:** Allows developers to define routing criteria

Example Use Cases

1. **Multilingual Chatbots:** Route queries to language-specific processing chains.
2. **Content Classification:** Direct content to specialized analysis chains based on topic.
3. **Task-Specific Routing:** Send user requests to the most suitable tool or agent.

CHAPTER 2   CORE COMPONENTS OF LANGCHAIN

Creating Route Templates

Example Code Implementation

Consider a scenario where we would wish to route it to different LLMs/Agents based on specific needs or scenarios.

```
# Templates for different user types
beginner_template = '''You are a physics teacher focused on explaining
complex topics in simple terms for beginners.
You assume no prior knowledge. Here is the question:\n{input}'''

expert_template = '''You are an expert physics professor who explains
topics to advanced audiences.
Assume the person has a PhD-level understanding. Here is the question:\
n{input}'''
```

Defining Prompt Information

```
# Define prompt information for routing
prompt_infos = [
    {'name':'advanced physics', 'description': 'Answers advanced physics
    questions', 'prompt_template':expert_template},
    {'name':'beginner physics', 'description': 'Answers basic beginner
    physics questions', 'prompt_template':beginner_template}
]
```

Creating Destination Chains: Next, we convert each prompt into an LLM chain and store them in a dictionary, which will be used for routing.

```
from langchain.chat_models import ChatOpenAI
from langchain.prompts import ChatPromptTemplate
from langchain.chains import LLMChain

# Initialize the LLM (Azure OpenAI in this case)
llm = AzureOpenAI(deployment_name="<Azure deployment name>", model_
name="<Model name>")

# Create chains for each destination
destination_chains = {}
for p_info in prompt_infos:
```

```
    name = p_info["name"]
    prompt_template = p_info["prompt_template"]
    prompt = ChatPromptTemplate.from_template(template=prompt_template)
    chain = LLMChain(llm=llm, prompt=prompt)
    destination_chains[name] = chain
```

Multi-prompt Routing

LangChain's MultiPromptChain allows us to dynamically choose the appropriate chain based on the user input.

```
from langchain.chains.router.multi_prompt_prompt import MULTI_PROMPT_ROUTER_TEMPLATE

# Join all destinations into a single string
destinations = [f"{p['name']}: {p['description']}" for p in prompt_infos]
destinations_str = "\n".join(destinations)

# Define the router template using the multi-prompt router
from langchain.prompts import PromptTemplate
router_template = MULTI_PROMPT_ROUTER_TEMPLATE.format(destinations=destinations_str)
router_prompt = PromptTemplate(template=router_template, input_variables=["input"])
```

This code snippet demonstrates LangChain's routing functionality using the Multi-prompt Router. It takes a list of destination prompts with their descriptions, formats them into a routing template, and creates a PromptTemplate that can direct user inputs to the most appropriate specialized prompt handler.

Routing Chain Implementation

We can now define the routing chain and connect the different destination chains to it. We also define a default chain for unmatched queries.

```
from langchain.chains.router.llm_router import LLMRouterChain, RouterOutputParser

# Set up the router chain
router_chain = LLMRouterChain.from_llm(llm, router_prompt)
```

CHAPTER 2 CORE COMPONENTS OF LANGCHAIN

```
# Define a default chain for unmatched queries
default_prompt = ChatPromptTemplate.from_template("{input}")
default_chain = LLMChain(llm=llm, prompt=default_prompt)

from langchain.chains.router import MultiPromptChain
chain = MultiPromptChain(router_chain=router_chain,
                        destination_chains=destination_chains,
                        default_chain=default_chain, verbose=True)

# Run the routing chain for different queries
chain.run("How do magnets work?")
Beginner physics: {'input': 'How do magnets work?'}
> Finished chain.
'\n\nMagnets work by creating a magnetic field around them. This magnetic
field is made up of tiny particles called electrons, which have a negative
charge. When two magnets are brought close together, the negative electrons
in one magnet are attracted to the positive electrons in the other
magnet.  I hope this explanation helps you understand how magnets work!'

chain.run("How do Feynman Diagrams work?")
Advanced physics: {'input': 'How do Feynman Diagrams work?'}
> Finished chain.
'\n\nFeynman diagrams are graphical representations of mathematical
equations that describe the behavior of particles in quantum field
theory. They were developed by physicist Richard Feynman. The rules for
constructing Feynman diagrams dictate that the total number of incoming
lines must equal the total number of outgoing lines at each vertex, and the
conservation laws of energy
```

The code implements a router chain system that intelligently directs queries to appropriate specialized chains based on complexity. It uses LLMRouterChain to analyze queries and route them to either beginner or advanced physics chains, with a default chain as fallback. The MultiPromptChain orchestrates this routing mechanism, ensuring queries receive appropriate-level responses.

RunnablePassthrough is a simple utility that acts as a "pass-through" component in a chain. It takes input and passes it forward to the next component in the chain without modifying it. Essentially, it serves as a placeholder or connector in a sequence of operations.

Steps used in the example above:

Template Definition: The code starts by defining two templates: beginner_template and expert_template. These templates are strings that will be used to format questions for different levels of expertise.

Prompt Information: A list called prompt_infos is created, containing dictionaries with information about different types of prompts. Each dictionary includes a name, description, and the corresponding template to use.

Creating Destination Chains: The code then iterates through the prompt_infos list to create a dictionary of "destination chains." Each chain is associated with a specific prompt type (e.g., "advanced physics" or "beginner physics"). These chains use the LLMChain class, which combines a language model (LLM) with a prompt template.

Default Chain: A default chain is created using a simple template that just passes the input directly. This will be used if no specific destination chain matches the input.

Formatting Destinations: The code creates a formatted string of destinations by combining the name and description of each prompt info.

Router Setup: A router template is created using a predefined MULTI_PROMPT_ROUTER_TEMPLATE and the formatted destinations. This template is used to create a router prompt, which will help decide which destination chain to use for a given input.

Creating the Router Chain: An LLMRouterChain is created using the language model and the router prompt. This chain will be responsible for routing inputs to the appropriate destination chain.

Multi-prompt Chain: Finally, a MultiPromptChain is created, combining the router chain, destination chains, and default chain. This setup allows the system to route questions to the most appropriate chain based on the input.

Routing by Semantic Similarity

Sometimes, routing based on predefined templates might not be enough. You can route queries based on semantic similarity using embeddings and cosine similarity.

```python
from langchain.utils.math import cosine_similarity
from langchain_core.output_parsers import StrOutputParser
from langchain_openai import AzureOpenAIEmbeddings

# Example templates for routing by similarity
physics_template = """You are a smart physics professor, great at answering questions about physics concisely.
When you don't know the answer, admit that you don't know."""

math_template = """You are a mathematician. You are great at breaking down complex problems and explaining step-by-step."""

# Embed the templates for comparison
embeddings = AzureOpenAIEmbeddings(model=text-embedding-ada-002")
prompt_templates = [physics_template, math_template]
prompt_embeddings = embeddings.embed_documents(prompt_templates)

# Function to route based on similarity
def prompt_router(input):
    query_embedding = embeddings.embed_query(input["query"])
    similarity = cosine_similarity([query_embedding], prompt_embeddings)[0]
    most_similar = prompt_templates[similarity.argmax()]
    print("Using MATH" if most_similar == math_template else "Using PHYSICS")
    return PromptTemplate.from_template(most_similar)

# Creating the chain
from langchain_core.runnables import RunnableLambda, RunnablePassthrough
chain = (
    {"query": RunnablePassthrough()}
    | RunnableLambda(prompt_router)
    | llm
    | StrOutputParser()
)
```

CHAPTER 2 CORE COMPONENTS OF LANGCHAIN

```
# Test the chain
print(chain.invoke("What's a black hole"))
Using PHYSICS
?

A black hole is a region in space where the gravitational force is so
strong that not even light can escape from it. This occurs when a massive
star collapses
print(chain.invoke("What is chain rule in differentiation?"))
Using MATH
The chain rule in differentiation is a method used to find the derivative
of a function that is composed of two or more functions.In other words, if
f(x) = g(h(x)), then the derivative of f(x) is equal to g'(h(x)) * h'(x).
```

The code implements a smart routing system using embeddings and cosine similarity. It converts user queries into embeddings, compares them with pre-embedded templates using cosine similarity, and routes to the most relevant expert template (math or physics). This ensures questions are answered by the most appropriate AI persona.

TransformChain: Data Transformation and Preprocessing

TransformChain is a specialized chain in LangChain designed for data transformation and preprocessing tasks. This chain type is crucial for preparing inputs before they are passed to other chains or language models, ensuring that the data is in the most suitable format for subsequent processing.

Key Features of TransformChain

- **Flexible Transformations:** Can perform various data manipulations
- **Input-Output Mapping:** Clearly defines how inputs are transformed into outputs
- **Chainable:** Can be easily integrated into larger chain sequences

Common Transformation Tasks

1. **Text Normalization:** Standardizing text format, case, or encoding
2. **Feature Extraction:** Deriving relevant features from raw input

3. **Data Cleaning:** Removing noise or irrelevant information
4. **Format Conversion:** Transforming data between different structures or schemas

Implementation

```python
from langchain.chains import TransformChain

def transform_func(inputs):
    text = inputs["text"]
    transformed_text = text.lower()  # Simple lowercase transformation
    word_count = len(text.split())
    return {"lowercase_text": transformed_text, "word_count": word_count}

transform_chain = TransformChain(
    input_variables=["text"],
    output_variables=["lowercase_text", "word_count"],
    transform=transform_func
)

result = transform_chain.invoke("This is an EXAMPLE sentence.")
print(result)
# Output: {'lowercase_text': 'this is an example sentence.', 'word_count': 5}
```

TransformChain allows developers to create clean, modular data preprocessing steps that can significantly improve the performance and reliability of their LangChain applications.

MapReduceChain: Parallel Processing for Large Datasets

MapReduceChain is an advanced chain type in LangChain that enables efficient processing of large datasets through parallel computation. This chain implements the MapReduce programming model, allowing developers to break down complex tasks into smaller, manageable pieces that can be processed concurrently.

Chapter 2 Core Components of LangChain

Key Features of MapReduceChain

- **Scalability:** Handles large volumes of data by distributing the workload
- **Parallel Processing:** Improves performance through concurrent execution
- **Flexible Mapping and Reducing:** Customizable functions for data processing
- **Aggregation:** Combines results from multiple parallel operations

The MapReduce process in LangChain typically involves three main steps:

1. **Map:** Apply a function to each item in the input dataset.
2. **Process:** Perform a specific operation on the mapped data (often using an LLM).
3. **Reduce:** Aggregate the results from the processed data.

Example Use Case

Summarizing a Large Document

For this example, please make use of this file from the repo: large_document.txt

```
from langchain.chains import LLMChain
from langchain.text_splitter import CharacterTextSplitter
from langchain.prompts import PromptTemplate
from langchain.llms import OpenAI
from langchain.chains.combine_documents.map_reduce import MapReduceDocumentsChain
from langchain.chains.combine_documents.stuff import StuffDocumentsChain
from langchain.docstore.document import Document

# Initialize the language model
llm = AzureOpenAI(deployment_name="<Azure deployment name>", model_name="<Model name>")

# Initialize text splitter
text_splitter = CharacterTextSplitter(
    separator="\n",
```

```python
    chunk_size=1000,
    chunk_overlap=200
)

# Define map and reduce prompts
map_prompt = PromptTemplate(
    input_variables=["text"],
    template="Summarize this text in one sentence: {text}"
)

reduce_prompt = PromptTemplate(
    input_variables=["text"],
    template="Combine these summaries into a coherent paragraph: {text}"
)

# Create the map and reduce chains
map_chain = LLMChain(llm=llm, prompt=map_prompt)
reduce_chain = LLMChain(llm=llm, prompt=reduce_prompt)

# Create combine documents chain
combine_documents_chain = StuffDocumentsChain(
    llm_chain=reduce_chain,
    document_variable_name="text"
)

# Create the map reduce chain
map_reduce_chain = MapReduceDocumentsChain(
    llm_chain=map_chain,
    # Use reduce_documents_chain instead of combine_documents_chain
    reduce_documents_chain=combine_documents_chain,
    document_variable_name="text",
    return_intermediate_steps=True
)

# Process large document
with open("large_document.txt", "r") as file:
    large_text = file.read()
```

CHAPTER 2   CORE COMPONENTS OF LANGCHAIN

```
# Split text into documents
texts = text_splitter.split_text(large_text)
docs = [Document(page_content=t) for t in texts]

# Process the documents
try:
    result = map_reduce_chain.invoke(docs)
    print("\nFinal Summary:")
    print(result['output_text'])
except Exception as e:
    print(f"An error occurred: {str(e)}")
```

output:
Final Summary:
Each paragraph should also flow logically from one to the next, building upon the previous one and contributing to the overall argument and thesis statement. Just as a seed needs a strong foundation and support to grow, a paper requires a strong argument and well-structured paragraphs to effectively convey its message. Brainstorming and organizing ideas are crucial steps in this process, as they help to develop a coherent and well-supported central idea. By utilizing different structures such as narration, description, process, classification, and illustration, a writer can effectively present and support their argument, just as each part of a seed supports its growth into a strong and thriving plant

MapReduceChain implements a divide-and-conquer approach for processing large texts. It first splits the text into manageable chunks, applies a summarization prompt to each chunk (map phase), and then combines these summaries into a final coherent result (reduce phase). This pattern is particularly effective for processing lengthy documents while maintaining context and coherence.

MapReduceChain is particularly useful for tasks such as document summarization, sentiment analysis of large datasets, or extracting key information from multiple sources. By leveraging parallel processing, it allows LangChain applications to handle much larger volumes of data than would be possible with sequential processing.

Custom Chains: Creating Specialized Chains for Specific Tasks

Custom chains in LangChain provide developers with the flexibility to create tailored solutions for specific tasks or workflows. By combining existing chain types, integrating custom logic, or implementing entirely new chain structures, developers can address unique requirements and optimize their language AI applications.

Key Advantages of Custom Chains

- **Task-Specific Optimization:** Design chains that perfectly fit the problem at hand.
- **Enhanced Control:** Fine-tune every aspect of the chain's behavior.
- **Improved Efficiency:** Streamline processes by eliminating unnecessary steps.
- **Innovation:** Implement novel approaches to language AI tasks.

Steps to Create a Custom Chain

1. Identify the specific requirements of your task.
2. Determine which existing chain types or components can be leveraged.
3. Design the structure and flow of your custom chain.
4. Implement custom logic or transformations as needed.
5. Integrate the custom chain with other LangChain components.

Example: Creating a custom chain for article generation with fact-checking.

```python
from langchain.chains import LLMChain, SequentialChain
from langchain.llms import AzureOpenAI
from langchain.prompts import PromptTemplate

class ArticleGeneratorChain(SequentialChain):
    def __init__(self, llm):
        # Article generation chain
        article_prompt = PromptTemplate(
            input_variables=["topic"],
            template="Write a short article about {topic}."
        )
```

```python
    article_chain = LLMChain(
        llm=llm,
        prompt=article_prompt,
        output_key="article"
    )

    # Fact-checking chain
    fact_check_prompt = PromptTemplate(
        input_variables=["article"],
        template="Identify any factual claims in this article and rate
        their accuracy: {article}"
    )
    fact_check_chain = LLMChain(
        llm=llm,
        prompt=fact_check_prompt,
        output_key="fact_check"
    )

    # Revision chain
    revision_prompt = PromptTemplate(
        input_variables=["article", "fact_check"],
        template="Revise this article based on the fact-check results:
        \n\n Article: {article}\n\nFact-check: {fact_check}\n\nRevised
        article:"
    )
    revision_chain = LLMChain(
        llm=llm,
        prompt=revision_prompt,
        output_key="revised_article"
    )

    # Initialize the sequential chain
    super().__init__(
        chains=[article_chain, fact_check_chain, revision_chain],
        input_variables=["topic"],
        output_variables=["article", "fact_check", "revised_article"]
    )
```

```
# Usage
llm = AzureOpenAI(deployment_name="<Azure deployment name>", model_
name="<Model name>")
article_generator = ArticleGeneratorChain(llm)
result = article_generator.invoke({"topic": "Singularity will be achieved
by 2025"})
print("fact_check results", result['fact_check'])
```

Output
```
fact_check results
 Singularity will eventually surpass human intelligence.
```

1. The concept of Singularity refers to a hypothetical future event in which artificial intelligence will surpass human intelligence.
Rating: Mostly accurate. While the concept of Singularity is still hypothetical, it is a widely recognized concept in the scientific and technological community.

2. Singularity is also known as the technological singularity.
Rating: Accurate.

3. Many experts and futurists have predicted that Singularity will be achieved by the year 2025.
Rating: Inaccurate. While some experts have made predictions about when Singularity may occur, there is no consensus on a specific timeframe.

This custom ArticleGeneratorChain combines multiple LLMChains to create a sophisticated workflow for generating articles with built-in fact-checking and revision. By creating custom chains, developers can tackle complex, multi-step tasks while maintaining the modularity and flexibility that LangChain offers.

Custom chains enable developers to push the boundaries of what's possible with language AI, creating innovative solutions tailored to their specific use cases and requirements.

Designing Effective Chains

In the world of large language models and AI applications, chains have become a crucial concept. They allow us to string together multiple operations or models to achieve complex tasks. This chapter probes into the art and science of designing effective chains, covering best practices, performance optimization, error handling, and debugging techniques.

Best Practices for Chain Design

When designing chains, several best practices can help ensure efficiency, maintainability, and scalability:

1. **Modular Design:** Break down complex tasks into smaller, reusable components. This approach enhances readability and allows for easier maintenance and updates.

2. **Clear Input and Output Definitions:** For each component in the chain, clearly define the expected input format and the output it produces. This clarity helps in debugging and ensures smooth data flow between components.

3. **Thoughtful Sequencing:** Arrange the components in a logical order that minimizes unnecessary processing and maximizes the utility of each step's output.

4. **Validate Inputs Early:** Implement input validation at the beginning of the chain to catch potential issues before they propagate through the system.

5. **Document Extensively:** Provide clear documentation for each component and the overall chain structure. This practice is crucial for team collaboration and future maintenance.

6. **Version Control:** Use version control systems to track changes in your chain designs, allowing for easy rollbacks and collaborative development.

7. **Consider Parallelization:** Where possible, design chains that can execute independent components in parallel to improve overall performance.

Optimizing Chain Performance

Performance optimization is crucial for ensuring that chains operate efficiently, especially when dealing with large-scale data or real-time applications. Here are some strategies for optimizing chain performance:

1. **Caching:** Implement caching mechanisms for frequently used or computationally expensive results. This can significantly reduce processing time for repeated operations.

   ```
   from langchain.cache import InMemoryCache
   from langchain.globals import set_llm_cache

   # Set up in-memory cache
   set_llm_cache(InMemoryCache())

   # Now any calls to LLMs will be cached
   ```

   Caching stores the results of expensive operations, such as LLM calls. When the same operation is requested again, the cached result is returned, saving time and resources. The code depends on both the langchain-core and langchain Python packages being installed in your environment. If these aren't installed, the code may fail due to missing dependencies.

2. **Lazy Evaluation:** Design chains to compute results only when they are needed, rather than eagerly processing all possible outcomes.

   ```
   # Lazy evaluation
   class LazyChain:
       def __init__(self, llm, prompt_template):
           self.llm = llm
           self.prompt_template = prompt_template  # Store as prompt_template
           self._result = None

       def get_result(self, input_text: str) -> str:
           if self._result is None:
               # Create PromptTemplate instance here
   ```

```
                prompt = PromptTemplate.from_template(self.prompt_
                template)
                chain = LLMChain(llm=self.llm, prompt=prompt)
                self._result = chain.run(input_text)
            return self._result

# Example usage
lazy_chain = LazyChain(llm, "Analyze: {text}")
result = lazy_chain.get_result("How is AI shaping the
modern world")
print("result:\n", result)
```

output:

Artificial intelligence (AI) is quickly shaping the modern world in a variety of ways. From revolutionizing industries to changing the way we live and work, AI has become an integral part of our daily lives. Here are some of the ways in which AI is shaping the modern world:

1. Automation and efficiency: One of the key ways in which AI is shaping the modern world is through automation and efficiency.

2. Personalization: AI is also shaping the modern world by personalizing experiences for individuals. This has led to improved customer satisfaction and loyalty.

3. Improving healthcare: AI is playing a crucial role in improving healthcare by accurately diagnosing diseases, assisting in surgeries, and developing personalized treatment plans.

This code implements lazy evaluation for a language model chain, deferring the actual computation until get_result() is called and caching the result to avoid redundant processing.

3. **Resource Allocation:** Carefully manage resources such as memory and CPU usage, especially for components that require significant computational power.

```
from langchain.chat_models import ChatOpenAI

llm = ChatOpenAI(
    max_tokens=1000,  # Limit token usage
    request_timeout=30  # Set timeout for requests
)
```

4. **Data Pipelining:** Implement data pipelining techniques to start processing subsequent steps before the entire dataset has been processed by earlier steps.

5. **Batch Processing:** Where applicable, process data in batches to take advantage of parallelization and reduce overhead.

6. **Optimize Individual Components:** Regularly profile and optimize the performance of individual components within the chain.

7. **Use Appropriate Data Structures:** Choose data structures that are best suited for the operations being performed in each component of the chain.

Error Handling and Robustness in Chains

Robust error handling is essential for creating reliable and maintainable chains. Consider the following approaches:

1. **Graceful Degradation:** Design chains to continue functioning, possibly with reduced capabilities, even when some components fail.

2. **Comprehensive Error Catching:** Implement try-catch blocks or equivalent error-handling mechanisms around critical sections of the chain.

3. **Detailed Error Logging:** Log errors with sufficient detail to facilitate debugging and monitoring.

4. **Retry Mechanisms:** Implement intelligent retry logic for transient failures, such as network issues or temporary service unavailability.

5. **Fallback Strategies:** Design alternative paths or default behaviors that the chain can follow when primary components fail.

6. **Input Sanitization:** Implement thorough input sanitization to prevent errors caused by unexpected or malformed inputs.

7. **State Management:** Implement robust state management to handle interruptions and allow for graceful recovery and resumption of chain execution.

Debugging and Testing Chains

Effective debugging and testing are crucial for maintaining and improving chain designs. Here are some strategies to consider:

1. **Unit Testing:** Develop comprehensive unit tests for individual components of the chain.

2. **Integration Testing:** Create tests that verify the correct interaction between different components in the chain.

3. **End-to-End Testing:** Implement tests that run through the entire chain to ensure overall functionality.

4. **Logging and Monitoring:** Implement detailed logging throughout the chain to track data flow and component performance.

5. **Visualization Tools:** Use visualization tools to represent the chain structure and data flow, aiding in understanding complex chains.

6. **Debugging Modes:** Implement debugging modes that can provide additional information or step-through capabilities during development.

7. **Mocking and Stubbing:** Use mocking frameworks to isolate and test individual components without dependencies on external services or data sources.

8. **Performance Profiling:** Regularly profile the performance of your chains to identify bottlenecks and optimization opportunities.

9. **Regression Testing:** Maintain a suite of regression tests to ensure that changes or optimizations don't introduce new bugs or reduce functionality.

By following these practices for design, optimization, error handling, and debugging, you can create robust, efficient, and maintainable chains that effectively leverage the power of language models and AI components.

Prompt Templates

Understanding the Importance of Prompts in LLMs

Large language models (LLMs) have revolutionized natural language processing, enabling a wide range of applications from chatbots to content generation. At the heart of effectively utilizing these models lies the art and science of crafting prompts. Prompts are the instructions or queries we provide to LLMs to elicit desired responses or behaviors.

The importance of prompts in LLMs cannot be overstated:

1. **Guidance:** Prompts serve as a guide, steering the model toward generating relevant and focused outputs.

2. **Context Setting:** They provide necessary context, helping the model understand the task at hand and produce appropriate responses.

3. **Task Specification:** Prompts define the specific task or query, whether it's answering a question, summarizing text, or generating creative content.

4. **Output Control:** Well-crafted prompts can influence the style, tone, and format of the model's output.

5. **Performance Optimization:** The quality of prompts directly impacts the performance of LLMs, affecting accuracy, relevance, and coherence of responses.

Understanding how to effectively use prompts is crucial for developers, researchers, and users of LLM-powered applications to harness the full potential of these powerful models.

Creating and Customizing Prompt Templates

Prompt templates are pre-designed structures for creating prompts that can be easily customized for specific use cases. They provide a consistent framework for interacting with LLMs while allowing flexibility in content.

Steps to create and customize prompt templates:

1. **Identify the Use Case:** Determine the specific task or application for which you're creating the template.

2. **Design the Basic Structure:** Create a skeleton that includes placeholders for variable content.

3. **Include Key Components:**
   - Task description
   - Context or background information
   - Specific instructions or constraints
   - Examples (if needed)
   - Output format specification

4. **Add Customization Points:** Identify areas where users can input specific details or modify the prompt.

5. **Test and Refine:** Try the template with various inputs and refine based on the results.

Example of a basic prompt template for a question-answering task:

```
Context: {context}
Question: {question}
Please provide a concise answer based on the given context. If the answer is not in the context, say "I don't have enough information to answer this question."
Answer:
In this template, {context} and {question} are customization points where specific content can be inserted.
```

This is a prompt template designed for question-answering, where placeholders {context} and {question} are dynamically replaced with specific text, allowing flexible extraction of answers from a given context.

Dynamic Prompt Generation

Dynamic prompt generation involves creating prompts on-the-fly based on user input, system state, or other variables. This approach allows for more flexible and context-aware interactions with LLMs.

Key aspects of dynamic prompt generation:

1. **Input Analysis:** Analyze user input or system variables to determine the appropriate prompt structure and content.

2. **Template Selection:** Choose the most suitable template based on the analyzed input.

3. **Variable Insertion:** Dynamically insert relevant information into the chosen template.

4. **Context Management:** Maintain and update context across multiple interactions when necessary.

5. **Adaptive Prompting:** Adjust prompts based on previous responses or user feedback.

Example of dynamic prompt generation in Python:

```python
def generate_dynamic_prompt(user_input, context):
    if "summarize" in user_input.lower():
        template = "Summarize the following text in 3 sentences: {text}"
        return template.format(text=context)
    elif "question" in user_input.lower():
        template = "Context: {context}\nQuestion: {question}\nAnswer:"
        return template.format(context=context, question=user_input)
    else:
        return f"Please respond to the following: {user_input}"
```

This function selects and populates different templates based on the user's input. The code demonstrates dynamic prompt generation by analyzing user intent through keywords. It adaptively selects and formats appropriate prompt templates based on whether the user wants a summary or asks a question. This approach enables more contextually relevant and structured interactions with the language model.

Prompt Optimization Techniques

Optimizing prompts is crucial for improving the performance and reliability of LLM outputs. Several techniques can be employed to enhance prompt effectiveness:

1. **Iterative Refinement:** Gradually improve prompts through multiple rounds of testing and adjustment.

2. **A/B Testing:** Compare different prompt variations to identify the most effective ones.

3. **Prompt Chaining:** Break complex tasks into a series of simpler prompts, using the output of one as input for the next.

4. **Few-Shot Learning:** Include relevant examples in the prompt to guide the model's understanding and output.

5. **Instruction Tuning:** Fine-tune the language model on a dataset of instructions and corresponding outputs.

6. **Prompt Ensembling:** Combine multiple prompts or prompt variations to improve robustness and accuracy.

7. **Temperature and Top-p Sampling Adjustment:** Experiment with different sampling parameters to balance creativity and coherence in outputs.

8. **Prompt Length Optimization:** Find the right balance between providing sufficient context and keeping the prompt concise.

9. **Error Analysis:** Systematically analyze model outputs to identify patterns in errors and refine prompts accordingly.

Example of prompt optimization through few-shot learning:

```python
from langchain.prompts import FewShotPromptTemplate, PromptTemplate

# Define the example prompt template
example_prompt = PromptTemplate(
    input_variables=["input", "output"],
    template="Input: {input}\nOutput: {output}"
)

# Define the few-shot prompt template
few_shot_prompt = FewShotPromptTemplate(
    examples=[
        {"input": "This film was a complete waste of time. The plot was
        confusing and the acting was terrible?", "output": "Negative"},
        {"input": "I was blown away by the stunning visuals and compelling
        storyline. A must-see!", "output": "Positive"},
    ],
    example_prompt=example_prompt,
    prefix="Answer the following questions:\n",
    suffix="{input}\nOutput:",
    input_variables=["input"],
)

# Use the few-shot prompt in your chain
chain = LLMChain(llm=llm, prompt=few_shot_prompt)
result = chain.invoke({"input": "very bad decision to watch that movie"})
print(result)
```

output:
{'input': 'very bad decision to watch that movie', 'text': ' Negative'}

This code demonstrates a few-shot prompt template for sentiment analysis using LangChain. It defines example inputs and outputs for positive and negative movie reviews, creating a prompt template that guides the language model to classify new movie review inputs based on these examples. The code sets up a chain that will predict the sentiment of a given movie review text.

CHAPTER 2   CORE COMPONENTS OF LANGCHAIN

Managing Prompt Libraries

As the use of LLMs in applications grows, managing a collection of prompts becomes increasingly important. A prompt library is an organized collection of prompts and templates that can be easily accessed, version-controlled, and maintained.

Key aspects of managing prompt libraries:

- **Categorization:** Organize prompts by task type, domain, or application for easy retrieval.

- **Version Control:** Use version control systems (e.g., Git) to track changes and maintain prompt history.

- **Documentation:** Include clear descriptions, usage instructions, and performance metrics for each prompt.

- **Standardization:** Establish naming conventions and formatting standards for consistency.

- **Testing Framework:** Implement automated testing to ensure prompt quality and performance across versions.

- **Collaborative Editing:** Use tools that allow team members to collaboratively develop and refine prompts.

- **Access Control:** Implement appropriate access controls to manage who can view, edit, or use different prompts.

- **Integration with Development Workflow:** Incorporate prompt management into the broader application development process.

Example structure for a prompt library:

```
prompt_library/
├── question_answering/
│   ├── general_qa.txt
│   ├── medical_qa.txt
│   └── legal_qa.txt
├── summarization/
│   ├── news_summary.txt
│   └── academic_paper_summary.txt
├── code_generation/
```

```
│   ├── python_function.txt
│   └── sql_query.txt
├── creative_writing/
│   ├── story_outline.txt
│   └── character_description.txt
└── README.md
```

By effectively managing a prompt library, organizations can maintain consistency, improve efficiency, and facilitate the sharing of best practices in working with LLMs.

Tools and Function Calling

Overview of Tools in LangChain

Tools in LangChain are fundamental components that enable large language models (LLMs) to interact with external systems and perform specific tasks. They serve as a bridge between the abstract reasoning capabilities of LLMs and concrete actions in the real world or within software environments.

At their core, tools in LangChain are Python functions wrapped with metadata. This metadata provides essential information about the tool, such as its name, description, and expected inputs. The LangChain framework uses this metadata to intelligently decide when and how to use each tool within a given context.

The primary purposes of tools in LangChain include

1. **Extending LLM Capabilities:** Tools allow LLMs to perform actions they couldn't do on their own, such as accessing real-time data, performing calculations, or interacting with APIs.

2. **Modularization:** By encapsulating specific functionalities into tools, developers can create more modular and maintainable code.

3. **Flexibility:** Tools can be easily swapped or combined to create complex chains of operations, allowing for highly flexible and customizable AI applications.

4. **Abstraction:** Tools provide a layer of abstraction between the LLM and the underlying implementation details, making it easier to work with diverse systems and data sources.

CHAPTER 2   CORE COMPONENTS OF LANGCHAIN

The LangChain framework provides a rich ecosystem of pre-built tools and the ability to create custom tools, making it a powerful platform for developing sophisticated AI applications.

Built-in Tools and Their Functionalities

LangChain offers a wide array of built-in tools that cover various common functionalities. These tools are designed to be easily integrated into AI applications with minimal setup. Here's an overview of some key categories of built-in tools and their functionalities:

1. **Search Tools**

    - **SerpAPI:** Enables web searches using the SerpAPI service
    - **GoogleSearchRun:** Performs Google searches and returns results
    - **WikipediaQueryRun:** Queries Wikipedia for information

2. **Data Manipulation Tools**

    - **PythonREPL:** Executes Python code and returns the output
    - **PythonAstREPLTool:** Similar to PythonREPL but uses AST for safer execution
    - **JsonListKeysTool:** Extracts keys from a JSON object

3. **File and System Tools**

    - **ReadFileTool:** Reads the content of a file
    - **WriteFileTool:** Writes content to a file
    - **ListDirectoryTool:** Lists files and directories in a given path

4. **Math and Calculation Tools**

    - **MathTool:** Performs mathematical calculations
    - **RequestsWrapper:** Sends HTTP requests and returns responses

5. **Language Processing Tools**

    - **TextRequestsWrapper:** Sends text-based HTTP requests
    - **QuerySQLDataBaseTool:** Executes SQL queries on a database

6. **API Integration Tools**

    - **OpenWeatherMapQueryRun:** Fetches weather data from OpenWeatherMap API
    - **ZapierNLARunAction:** Executes Zapier Natural Language Actions

7. **Vector Store Tools**

    - **VectorStoreQATool:** Performs question-answering tasks using a vector store
    - **VectorStoreQAWithSourcesTool:** Similar to VectorStoreQATool but includes source information

Usage: To use these built-in tools, you typically need to import them from the appropriate LangChain module and initialize them with any required parameters. For example:

```
from langchain_community.tools import DuckDuckGoSearchRun
search = DuckDuckGoSearchRun()
search.invoke("Einstein's first name?")
```

output:
"Albert Einstein, the brilliant physicist and Nobel laureate, revolutionized our understanding of the universe with his theory of relativity and became a symbol of genius that continues to inspire minds worldwide. Albert Einstein is one of..."

These tools can then be passed to an agent or used individually within your LangChain application.

CHAPTER 2   CORE COMPONENTS OF LANGCHAIN

Creating Custom Tools

While LangChain provides a comprehensive set of built-in tools, there are often scenarios where you need to create custom tools to meet specific requirements. Creating custom tools in LangChain is a straightforward process that involves defining a function and wrapping it with the appropriate tool decorator.

In Figure 2-2, let's see how we can create a custom tool step by step.

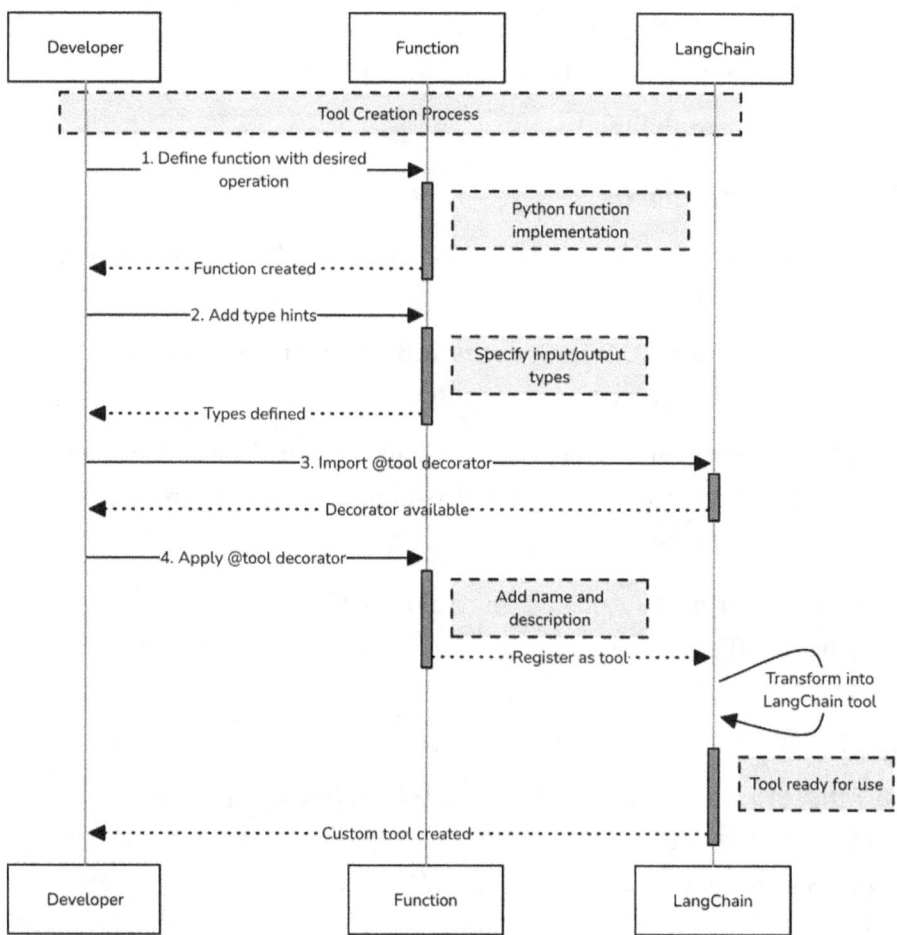

Figure 2-2. *Step-by-step process for creating a custom tool in LangChain*

1. **Define the Tool Function:** Create a Python function that performs the desired operation. This function should take input parameters and return a result.

2. **Add Type Hints:** Use Python type hints to specify the input and output types of your function. This helps LangChain understand how to use the tool.

3. **Use the @tool Decorator:** Import the "@tool" decorator from LangChain and apply it to your function. This decorator transforms your function into a LangChain tool.

4. **Provide Metadata:** In the decorator, specify the name and description of your tool. This information is used by LangChain to determine when and how to use the tool.

Here's an example of creating a custom tool that calculates the square of a number:

```python
from langchain.tools import tool
from typing import Union

@tool(name="SquareCalculator", description="Calculates the square of a given number")
def calculate_square(number: Union[int, float]) -> Union[int, float]:
    """
    Calculate the square of a given number.

    Args:
        number (Union[int, float]): The number to be squared.

    Returns:
        Union[int, float]: The square of the input number.
    """
    return number**2
```

output:
100.0

This code defines a LangChain tool called SquareCalculator that calculates the square of a number, using type hints to support both integer and float inputs, and providing a docstring explaining its functionality and parameters.

*****

For more complex tools, you can create a custom tool class by inheriting from "BaseTool":

```python
from langchain.tools import BaseTool
from pydantic import BaseModel, Field
from typing import Optional, Type

class FactorialInput(BaseModel):
    number: int = Field(..., description="The number to calculate the factorial of")

class FactorialTool(BaseTool):
    name: str = "FactorialCalculator"
    description: str = "Calculates the factorial of a given non-negative integer"
    args_schema: Optional[Type[BaseModel]] = FactorialInput

    def _run(self, number: int) -> int:
        if number < 0:
            raise ValueError("Factorial is only defined for non-negative integers")
        result = 1
        for i in range(1, number + 1):
            result *= i
        return result

    async def _arun(self, number: int) -> int:
        # For async execution
        return self._run(number)

tool = FactorialTool()
tool.run({"number": 5})
```

Output:
120

This code defines a LangChain tool FactorialTool that calculates the factorial of a non-negative integer, using Pydantic for input validation, with both synchronous and asynchronous execution methods, and implements error handling for negative inputs.

CHAPTER 2 CORE COMPONENTS OF LANGCHAIN

When creating custom tools, consider the following best practices:

- Provide clear and concise descriptions for your tools.
- Use type hints to ensure proper input validation.
- Handle potential errors and edge cases within your tool function.
- For complex tools, consider implementing both synchronous ("_run") and asynchronous ("_arun") versions.

Custom tools can be used alongside built-in tools in your LangChain applications, providing a powerful way to extend the capabilities of your AI systems.

Integrating External APIs As Tools

Integrating external APIs as tools in LangChain allows you to extend the capabilities of your AI applications by leveraging third-party services and data sources. This process involves creating a custom tool that wraps the API calls and handles the communication with the external service.

Figure 2-3 is a detailed guide on how to integrate an external API as a tool in LangChain.

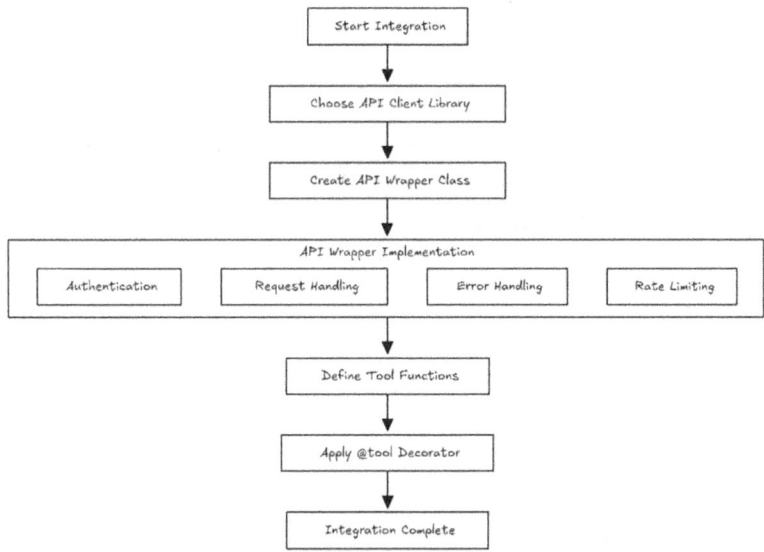

Figure 2-3. *Integrating external APIs as tool in LangChain*

49

1. **Choose an API Client Library:** Select an appropriate library for making HTTP requests. While you can use the built-in "requests" library, libraries like "aiohttp" for asynchronous requests or "httpx" for both synchronous and asynchronous requests are often preferred for their additional features and performance benefits.

2. **Create an API Wrapper Class:** Develop a Python class that encapsulates the API functionality. This class should handle authentication, request formation, and response parsing.

3. **Define Tool Functions:** Create methods within your API wrapper class that correspond to specific API endpoints or functionalities you want to expose as tools.

4. **Use the @tool Decorator:** Apply the "@tool" decorator to the methods you want to expose as tools, providing appropriate names and descriptions.

5. **Handle Errors and Rate Limiting:** Implement error handling and respect any rate limiting imposed by the API to ensure robust operation of your tools.

Here's an example of integrating a hypothetical weather API as a LangChain tool.

Note The URL `https://api.exampleweather.com/current` and the API key "your_api_key" are placeholders. Please replace them with your actual weather API endpoint and valid key from a real provider like OpenWeatherMap or WeatherAPI.

```
import httpx
from langchain.tools import BaseTool
from pydantic import BaseModel

class WeatherTool(BaseTool):
    name: str = "WeatherInfo"
    description: str = "Fetch current weather information for a specified city"
```

```python
def _run(self, city: str) -> str:
    """Fetch current weather information for a specified city."""
    try:
        url = "https://api.exampleweather.com/current"
        # Hypothetical URL
        params = {"city": city, "apiKey": "your_api_key"}
        # Assume API key is required
        response = httpx.get(url, params=params)
        response.raise_for_status()
        data = response.json()

        weather_condition = data["weather"]["condition"]
        temperature = data["weather"]["temperature"]
        return f"Current weather in {city}: {weather_condition}, 
        Temperature: {temperature}°C"
    except httpx.HTTPStatusError as e:
        return f"Error: Unable to fetch weather data. Status code: 
        {e.response.status_code}"
    except KeyError:
        return "Error: Unexpected response format from the weather API"
    except Exception as e:
        return f"An unexpected error occurred: {str(e)}"

async def _arun(self, city: str) -> str:
    """Asynchronous version of the weather fetching tool."""
    try:
        async with httpx.AsyncClient() as client:
            url = "https://api.exampleweather.com/current"
            params = {"city": city, "apiKey": "your_api_key"}
            response = await client.get(url, params=params)
            response.raise_for_status()
            data = response.json()

            weather_condition = data["weather"]["condition"]
            temperature = data["weather"]["temperature"]
            return f"Current weather in {city}: {weather_condition}, 
            Temperature: {temperature}°C"
```

```
        except httpx.HTTPStatusError as e:
            return f"Error: Unable to fetch weather data. Status code:
            {e.response.status_code}"
        except KeyError:
            return "Error: Unexpected response format from the weather API"
        except Exception as e:
            return f"An unexpected error occurred: {str(e)}"
```

To use this tool in your LangChain application:

```
weather_tool = WeatherTool(api_key="your_weather_api_key")
tools = [weather_tool]

# Use the tools in your agent or chain
agent = initialize_agent(tools, llm, agent=AgentType.ZERO_SHOT_REACT_
DESCRIPTION, verbose=True)
agent.run("What's the weather like in London?")
```

This code defines a WeatherTool class that fetches current weather information for a given city using an example weather API. It provides both synchronous (_run) and asynchronous (_arun) methods to retrieve weather data, handling potential errors like API access issues or unexpected response formats, and returning a formatted weather description.

When integrating external APIs, consider the following best practices:

- Use environment variables or secure secret management for API keys.

- Implement proper error handling and provide informative error messages.

- Consider implementing retries for transient failures.

- Use asynchronous methods ("_arun") for better performance in concurrent scenarios.

- Respect API rate limits and implement appropriate throttling mechanisms.

- Cache responses when appropriate to reduce API calls and improve performance.

By following these guidelines, you can create robust and efficient tools that seamlessly integrate external APIs into your LangChain applications.

Function Calling: Enhancing LLM Capabilities

Function calling is a powerful feature in LangChain that allows large language models (LLMs) to interact with predefined functions or tools in a more structured and controlled manner. This capability significantly enhances the LLM's ability to perform specific tasks, access external data, and generate more accurate and contextually relevant responses.

At its core, function calling in LangChain involves the following components:

1. **Function Definitions:** These are the specifications of functions that the LLM can call, including their names, descriptions, and expected parameters.

2. **LLM Integration:** The LLM is trained or fine-tuned to recognize when to use these functions and how to format the inputs correctly.

3. **Execution Environment:** A system that can interpret the LLM's function calls and execute the corresponding actual functions or tools.

Implementing Function Calling in LangChain

1. **Using LangChain's Tool System:** You can wrap your functions as LangChain tools and use them with agents.

    ```
    from langchain.agents import initialize_agent, Tool
    #  from langchain.llms import OpenAI

    def get_weather(location):
      return f"The weather in {location} is Summer."

    tools = [
       Tool(
           name="WeatherTool",
           func=get_weather,
           description="Useful for getting weather information for a specific location"
       )
    ]
    ```

```
llm = AzureOpenAI(deployment_name="<Azure deployment name>",
model_name="<Model name>")
agent = initialize_agent(tools, llm, agent="zero-shot-react-
description", verbose=True)

agent.run("What's the weather like in India?")
```

This code sets up a LangChain agent with a weather tool. The agent, powered by AzureOpenAI, can process a query about weather in a specific location by using the defined get_weather function. When asked about London's weather, it will invoke the tool to retrieve and potentially generate a weather-related response.

2. **Advanced Function Calling Techniques**
   Chained Function Calls:

   Implement complex workflows by chaining multiple function calls together.
```
from langchain.prompts import ChatPromptTemplate
from langchain.chains import LLMChain

llm = AzureOpenAI(deployment_name="<Azure deployment name>",
model_name="<Model name>")
prompt = ChatPromptTemplate.from_template(
    "Given the weather: {weather}, suggest an appropriate outfit."
)

weather_chain = LLMChain(llm=llm, prompt=prompt)
outfit_chain = LLMChain(llm=llm, prompt=ChatPromptTemplate.from_
template(
    "Describe the outfit: {outfit}"
))
```

```
def get_outfit_recommendation(location):
    weather = get_weather(location)
    outfit = weather_chain.run(weather=weather)
    return outfit_chain.run(outfit=outfit)

result = get_outfit_recommendation("India")
result
```

output:
Bright colors and bold patterns are popular in Indian fashion, so don't be afraid to embrace them!

The first chain (weather_chain) determines an outfit given weather conditions, and the second chain (outfit_chain) describes the recommended outfit in more detail. The get_outfit_recommendation function orchestrates this process by first fetching weather data, then generating and refining an outfit suggestion.

3. **Dynamic Function Generation:** Generate functions dynamically based on the conversation context or user input.

```
import json
from langchain.prompts import PromptTemplate
from langchain.chains import LLMChain

llm = AzureOpenAI(deployment_name="<Azure deployment name>",
model_name="<Model name>")
function_generator_prompt = PromptTemplate(
    input_variables=["context"],
    template="Based on the following context, generate a JSON specification for a function that would be useful: {context}\nEnsure the output is a single, valid JSON object." # Added instruction for valid JSON
)

function_generator_chain = LLMChain(llm=llm, prompt=function_generator_prompt)
```

```python
    def generate_dynamic_function(context):
        function_spec = function_generator_chain.run(context=context)
        try:
            # Attempt to parse JSON, handle errors gracefully
            return json.loads(function_spec)
        except json.JSONDecodeError as e:
            print(f"Error decoding JSON: {e}\nRaw output: {function_
            spec}")  # Print error and raw output for debugging
            return None  # Or handle the error differently

context = "The user is asking about financial calculations related
to mortgages."
dynamic_function = generate_dynamic_function(context)

Output
{'functionName': 'calculateMortgage',
 'description': 'This function calculates various financial values
related to mortgages.',
 'parameters': [{'name': 'principal',
    ...........
    principal and interest.'}}}
```

This code integrates LangChain with LLM to dynamically generate a JSON specification for a function based on a given context. It uses a PromptTemplate to instruct the LLM to produce a valid JSON object and ensures robustness by attempting to parse the JSON output, handling errors gracefully if the output is invalid. The example context provided relates to financial calculations for mortgages.

4. **Meta-Learning for Function Calling:** Implement a system where the LLM learns to improve its function-calling abilities over time.

```python
from langchain.memory import ConversationBufferMemory
from langchain.chains import ConversationChain
from langchain.chat_models import ChatOpenAI

# Assuming 'chat' is your ChatOpenAI instance
```

CHAPTER 2　CORE COMPONENTS OF LANGCHAIN

```python
llm = AzureOpenAI(deployment_name="<Azure deployment name>",
model_name="<Model name>")

memory = ConversationBufferMemory()
conversation = ConversationChain(
    llm=llm,
    memory=memory,
    verbose=True
)

def meta_function_caller(query, available_functions):
    # 1. Provide context and ask for function choice
    conversation.predict(
        input=f"Query: {query}\nAvailable functions:
        {', '.join(available_functions)}"
    )
    # 2. Get the function choice (extract from response if needed)
    function_choice_response = conversation.predict(input="Which
    function should be called for this query?")
    function_choice = function_choice_response.strip() # Extract
    and clean the function name
    # You might need more sophisticated logic to extract the
    function name reliably

    # 3. Validate function choice (optional but recommended)
    if function_choice not in available_functions:
        print (f"Warning: Invalid function choice '{function_
        choice}'.")
        # Handle invalid choice, e.g., reprompt or choose
        a default

    return function_choice

query = "What is the stock that I can buy for $1000 over 5 years
at 5% annual rate?"
available_functions = ["calculate_compound_interest", "get_stock_
price", "convert_currency"]
chosen_function = meta_function_caller(query, available_functions)
```

57

CHAPTER 2   CORE COMPONENTS OF LANGCHAIN

This code creates a conversational AI system using LangChain's ConversationChain with memory. The meta_function_caller method analyzes a query, consults the language model to select an appropriate function from available options, and returns the chosen function name for potential execution.

5. **Multi-step Reasoning:** Implement a system where the LLM can break down complex queries into multiple function calls, chaining them together to arrive at a final result.

```python
from langchain.prompts import PromptTemplate
from langchain.chains import LLMChain

reasoning_prompt = PromptTemplate(
    input_variables=["query", "available_functions"],
    template="Query: {query}\nAvailable functions: {available_functions}\nBreak down the steps needed to answer this query using the available functions."
)

reasoning_chain = LLMChain(llm=llm, prompt=reasoning_prompt)

def multi_step_function_calling(query, available_functions):
    steps = reasoning_chain.run(query=query, available_functions=available_functions)
    # Implement logic to execute each step and
    aggregate results
    return steps

query = "What's the price difference between Apple and Microsoft stocks over the last month?"
available_functions = ["get_stock_price", "calculate_percentage_change", "get_date_range"]
execution_plan = multi_step_function_calling(query, available_functions)
```

output:

Step 1: Use the get_date_range function to get the date range of the last month.

Step 2: Use the get_stock_price function to get the stock prices of Apple and Microsoft on the starting and ending dates of the last month.

Step 3: Calculate the percentage change of Apple and Microsoft stocks using the calculate_percentage_change function.

Step 4: Subtract the percentage change of Apple from the percentage change of Microsoft to find the price difference between the two stocks.

Step 5: Print or display the result to the user.

This code creates a multi-step function calling mechanism using Langchain. The multi_step_function_calling method uses an LLM chain to break down a query into executable steps by analyzing available functions, generating a potential execution plan for complex queries like comparing stock prices.

Best Practices for Function Calling in LangChain

- **Clear Function Descriptions:** Provide detailed and unambiguous descriptions for each function to help the LLM understand when and how to use them.

- **Type Checking and Validation:** Implement strict type checking and input validation for function parameters to prevent errors and unexpected behavior.

- **Graceful Error Handling:** Design your functions to handle potential errors gracefully and provide informative error messages that the LLM can interpret and act upon.

- **Versioning:** Implement a versioning system for your functions to manage changes over time and ensure compatibility with different LLM versions.

- **Monitoring and Logging:** Set up comprehensive monitoring and logging for function calls to track usage patterns, identify potential issues, and optimize performance.

- **Security Considerations:** Implement proper security measures, such as input sanitization and access controls, especially when dealing with sensitive operations or data.

Continuous Improvement: Regularly analyze the LLM's function-calling behavior and use insights to refine function definitions and improve overall system performance.

In this chapter, we explored the core building blocks of LangChain applicationsChains, Prompt Templates, and Tools. You learned how to structure workflows using chains, craft effective prompts, and integrate both built-in and custom tools with intelligent selection logic. These components form the foundation of any powerful LangChain based system.

Key Takeaways

- Chains are the fundamental building blocks in LangChain that enable sequential logic with LLMs.

- Prompt templates allow for the reuse and parameterization of LLM prompts across applications.

- Combining chains with prompt templates helps build modular, maintainable, and scalable LLM-powered workflows.

In the next chapter, we shift our focus to more advanced components that enable real-world, production-grade AI systems. You'll explore output parsers to structure model responses, memory modules for retaining context across interactions, and embeddings with vector stores for powerful semantic search.

CHAPTER 3

Advanced Components and Integrations

While Chapter 2 equipped you with foundational knowledge of chains, tools, and prompt templates, this chapter delves into the sophisticated building blocks that enable the creation of truly dynamic, intelligent, and scalable AI applications.

This chapter begins by exploring **output parsers**, essential tools for structuring the often-unstructured outputs of language models into actionable formats. We then transition to **memory components**, which empower your applications with context retention for more nuanced interactions. The use of **embeddings and vector stores** for semantic search and similarity matching highlights LangChain's capabilities in handling large datasets and enabling context-aware responses.

Next, we introduce **agents**, the decision-makers that combine tools, chains, and logic to autonomously solve complex tasks. The **callback system** is another key feature that provides powerful hooks for debugging, monitoring, and customizing chain executions. Finally, we will examine **chat models** and their unique differences from standard language models, as well as the **LangChain Expression Language (LCEL)**, a versatile tool for advanced chain configurations.

By the end of this chapter, you'll not only understand these advanced components but also learn how to integrate them seamlessly into your workflows, enabling you to build robust, production-grade applications that leverage the full power of LangChain. Whether you're enhancing conversational agents, optimizing workflows, or scaling AI solutions, the skills and insights from this chapter will unlock new possibilities for your AI-driven projects.

© Sanath Raj B Narayan and Nitin Agarwal 2025
S. R. B. Narayan and N. Agarwal, *Mastering LangChain*, https://doi.org/10.1007/979-8-8688-1718-2_3

CHAPTER 3　ADVANCED COMPONENTS AND INTEGRATIONS

Output Parser

An output parser is responsible for taking the output generated by a language model (LLM) and transforming it into a format that is more suitable for downstream tasks. This functionality is particularly useful when using LLMs to generate structured data or to normalize outputs from chat models and LLMs.

Output parsers are essential for

- **Generating Structured Data:** Extracting structured outputs like JSON, XML, or CSV from the unstructured responses of LLMs
- **Normalizing Outputs:** Ensuring consistency in outputs from various LLMs or chat models
- **Error Handling:** Wrapping around other parsers to fix or retry when errors occur in parsing

Types of Output Parsing

1. Str Parser

    - **Supports Streaming:** Yes.
    - **Use Case:** Ideal for extracting raw text from varied message formats. For instance, extracting relevant portions from a conversational chat log.

2. JSON Parser

    - **Supports Streaming:** Yes.
    - **Has Format Instructions:** Yes.
    - **Use Case:** Best for generating structured data in JSON format. Commonly used when integrating with APIs or storing data in NoSQL databases.

3. XML Parser

    - **Supports Streaming:** Yes.
    - **Has Format Instructions:** Yes.
    - **Use Case:** Use when working with models that need to produce XML-formatted outputs, such as for legacy systems or specific integrations.

CHAPTER 3   ADVANCED COMPONENTS AND INTEGRATIONS

4. CSV Parser

    - **Supports Streaming:** Yes.

    - **Has Format Instructions:** Yes.

    - **Use Case:** Excellent for handling tabular data or preparing outputs for spreadsheet software.

5. OutputFixing Parser

    - **Calls LLM:** Yes.

    - **Use Case:** Provides error handling by wrapping another parser. If the original parser encounters errors, it prompts the LLM to correct the output.

6. RetryWithError Parser

    - **Calls LLM:** Yes.

    - **Use Case:** Similar to OutputFixingParser but includes additional context like original inputs and error messages to improve the retry process.

7. Pydantic Parser

    - **Has Format Instructions:** Yes.

    - **Use Case:** Converts outputs into a Pydantic model, making it ideal for applications requiring strict data validation.

8. YAML Parser

    - **Has Format Instructions:** Yes.

    - **Use Case:** Similar to Pydantic but encodes the output in YAML, useful for configuration files or scenarios where YAML is the preferred format.

Choosing the Right Parser

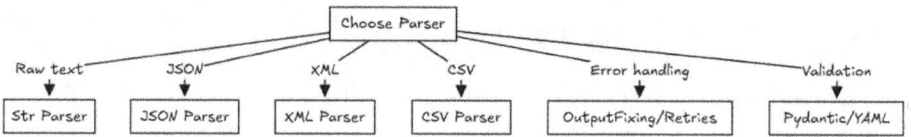

Figure 3-1. *Selecting the appropriate parser*

Selecting the appropriate parser (Figure 3-1) depends on the specific requirements of your application:

- **If You Need Raw Text:** Use the **Str** parser.
- **For Structured JSON Data:** Opt for the **JSON** parser.
- **For XML Outputs:** Use the **XML** parser.
- **For Tabular Data:** Choose the **CSV** parser.
- **For Robust Error Handling:** Use **OutputFixing** or **RetryWithError** parsers.
- **For Strict Data Validation:** Opt for the **Pydantic** or **YAML** parsers.

Output parsers in LangChain are powerful tools that streamline the process of transforming unstructured LLM outputs into structured, actionable formats. By understanding their features and use cases, you can effectively integrate them into your workflows to enhance data reliability and application performance.

Structured Output Parsing

Structured output parsing in LangChain is a powerful feature that allows developers to transform raw outputs from large language models (LLMs) into structured formats, such as JSON or Python dictionaries. This capability is essential for applications that require organized data for further processing, storage, or analysis.

Key Components of Structured Output Parsing

- **StructuredOutputParser:** Designed to extract information from LLM responses according to a predefined schema. Developers can define a ResponseSchema, which specifies the expected keys and their corresponding data types, ensuring that the model's output adheres to a specific structure. This is particularly useful when the output contains multiple fields or when the built-in parsers do not meet specific structural requirements.

- **PydanticOutputParser:** Leverages Pydantic models to enforce type safety and validation on the LLM's responses. This parser ensures that the output conforms strictly to the defined model schema, making it suitable for applications where data integrity is crucial.

- **Schema Definition:** The schema defines how the output should be structured. Each schema consists of key-value pairs, where each key represents a field name and has an associated description and expected data type (e.g., string, integer). This structured approach facilitates easier parsing and manipulation of the data returned by the model.

Practical Applications

- **Data Extraction:** When you need to extract specific information from an LLM's response—such as names, dates, or other entities—the StructuredOutputParser can be configured to parse these elements into a structured format that can be easily used in applications like databases or APIs.

- **Custom Formatting:** Developers can use structured output parsing to instruct LLMs on how to format their responses. This is achieved through methods like get_format_instructions(), which provide guidelines for structuring the output appropriately.

- **Complex Workflows:** By integrating structured output parsing into complex workflows, developers can build robust applications that handle intricate tasks while ensuring that outputs are organized and usable.

CHAPTER 3   ADVANCED COMPONENTS AND INTEGRATIONS

Example: Chatbot applications; maintaining a consistent response format enhances user experience and facilitates data handling.

```python
from langchain.output_parsers import StructuredOutputParser, ResponseSchema
from langchain_openai import AzureOpenAI
from langchain import PromptTemplate, LLMChain

# Define response schemas
response_schemas = [
    ResponseSchema(name="answer", description="Answer to the user's question"),
    ResponseSchema(name="fact", description="An interesting fact about the answer")
]

# Initialize the parser
output_parser = StructuredOutputParser.from_response_schemas(response_schemas)

# Get formatting instructions for prompts
format_instructions = output_parser.get_format_instructions()

# Example usage with an LLM
llm = AzureOpenAI(deployment_name="<LLM Deployment Name>", model_name="<Model Name>")

# Create a prompt template
prompt_template = PromptTemplate(
    template="What is the powerhouse of the cell? {format_instructions}",
    input_variables=["format_instructions"],
)
prompt = prompt_template.format(format_instructions=format_instructions)

# Create an LLMChain to integrate the parser
chain = LLMChain(llm=llm, prompt=prompt_template, output_parser=output_parser)

# Run the chain to get structured output
structured_output = chain.invoke(input=format_instructions)
print(structured_output)
```

output:
```
{'answer': 'The powerhouse of the cell is the mitochondria.', 'fact':
'Mitochondria have their own DNA and can reproduce independently within
a cell.'}
```

The code defines a LangChain pipeline to query an LLM (AzureOpenAI) for a structured response. It uses a StructuredOutputParser with predefined schemas to parse the LLM's output.

Custom Output Parsers for Specific Data Formats

Custom output parsers in LangChain allow developers to handle specific data formats and tailor the language model's outputs to meet the requirements of their application. By implementing custom parsers, you can define how the model's raw text output should be processed, validated, and transformed into usable data structures like JSON, lists, or domain-specific formats.

When to Use Custom Output Parsers

- When the model's output needs to conform to a specific schema or data structure

- To handle complex or domain-specific formats, such as SQL queries, JSON payloads, or structured CSV data

- To validate and clean the model's output for use in downstream applications or APIs

Creating a Custom Output Parser

To create a custom output parser in LangChain, implement the BaseOutputParser class or subclass an existing parser, overriding methods like parse and get_format_instructions.

Key Methods in a Custom Parser

- parse: Defines how the raw output is processed into the desired format

- get_format_instructions: Provides the instructions that the language model should follow to produce outputs in the expected format

Chapter 3   Advanced Components and Integrations

Example 1: Parsing JSON data

```python
from langchain.schema import BaseOutputParser
import json

class JSONOutputParser(BaseOutputParser):
    def parse(self, text: str):
        try:
            return json.loads(text)
        except json.JSONDecodeError as e:
            raise ValueError(f"Failed to parse JSON: {e}")

    def get_format_instructions(self) -> str:
        return "Provide the output in a valid JSON format."

# Usage
parser = JSONOutputParser()

# Example raw output
model_output = '{"name": "Alice", "age": 30}'
parsed_data = parser.parse(model_output)
print(parsed_data)
```

output:
{'name': 'Alice', 'age': 30}

This code defines a custom output parser JSONOutputParser that extends LangChain's BaseOutputParser, which attempts to parse raw text into JSON format and raises an error if parsing fails. It also includes a method for providing format instructions to the user, ensuring the output is in valid JSON format.

Example 2: Handling SQL Queries

```python
class SQLParser(BaseOutputParser):
    def parse(self, text: str):
        if not text.lower().startswith("select"):
            raise ValueError("Invalid SQL query. Expected a SELECT statement.")
        return text.strip()
```

```
    def get_format_instructions(self) -> str:
        return "Generate a valid SQL SELECT query."

# Usage
parser = SQLParser()

model_output = "SELECT * FROM users WHERE age > 25;"  # Example raw output
parsed_sql = parser.parse(model_output)
print(parsed_sql)
```

The above code has SQLParser class that extends BaseOutputParser to validate and parse SQL queries, ensuring they begin with "SELECT". The parse method checks the query format, while the get_format_instructions method provides instructions for generating a valid SQL SELECT query.

Error Handling in Output Parsing

When working with language models, it's common to encounter unexpected outputs or formatting issues that can disrupt the flow of your application. LangChain provides mechanisms for handling such errors gracefully, ensuring that your workflows are robust and resilient to parsing failures.

Common Issues in Output Parsing

- **Invalid Formats:** Model outputs don't adhere to the expected schema (e.g., malformed JSON or incorrect syntax).
- **Incomplete Outputs:** Truncated responses due to token limits or other issues.
- **Ambiguity:** Outputs are unclear or ambiguous, making parsing unreliable.
- **Logical Errors:** Outputs that are syntactically correct but don't meet the application's logical requirements.

Structured Output Validation
Use tools like Pydantic to define schemas and validate outputs. Raise an error when outputs don't match the schema.

Best Practices for Error Handling in Output Parsing

- **Precise Prompts:** Craft prompts that explicitly instruct the model to produce outputs in a specific format, reducing the likelihood of parsing errors.

- **Extensive Testing:** Test your parser against a variety of model outputs, including edge cases.

- **Graceful Degradation:** Design parsers and chains to handle errors without disrupting the user experience.

- **Use Validation Libraries:** Leverage tools like Pydantic for schema validation to ensure structured outputs.

- **Monitor and Iterate:** Continuously monitor parsing errors in production and iterate on prompts and parsers to improve robustness.

```
from langchain.schema import BaseOutputParser
from langchain.schema.output_parser import OutputParserException

class MyCustomParser(BaseOutputParser):
    def parse(self, text: str) -> str:
        if not isinstance(text, str):
            raise OutputParserException("Expected string input")

        if not text.startswith("Result:"):
            raise OutputParserException("Output must start with 'Result:'")

        return text.split("Result:", 1)[1].strip()

# Example usage
parser = MyCustomParser()
try:
    parsed_output = parser.parse("Result: This is valid output")
    print(f"Success: {parsed_output}")
except OutputParserException as e:
    print(f"Error: {e}")
```

This code defines a custom output parser class MyCustomParser that inherits from BaseOutputParser and implements the parse method to process text input. The parser checks if the input is a string starting with "Result:" and extracts the content following it, raising exceptions if these conditions aren't met.

Integrating Parsers with Chains and Models

Integrating output parsers with LangChain's chains and models enables the seamless conversion of language model outputs into structured, usable data. This integration helps automate workflows, validates the outputs for specific use cases, and ensures consistency across applications.

- **Structured Outputs:** Convert raw model outputs into structured data formats such as JSON, dictionaries, or domain-specific formats.
- **Validation:** Ensure the outputs conform to the required schema or structure.
- **Automation:** Simplify downstream processing by handling output parsing automatically.
- **Error Handling:** Build robust pipelines with fallback mechanisms when outputs are invalid.

Steps to Integrate Parsers with Chains

1. Define the parser.
2. Implement a parser by extending the BaseOutputParser class or using predefined parsers like PydanticOutputParser.
3. Provide format instructions.
4. Parsers often include a get_format_instructions method that tells the language model how to structure its output.
5. Integrate with PromptTemplate.
6. Use the parser's format instructions in the prompt to guide the model to produce outputs in the desired format.
7. Use the chain.
8. Pass the parser to the chain to automatically parse and validate the model's outputs.

Code Implementation

```python
# With LCEL
from langchain.prompts import ChatPromptTemplate
from langchain.chat_models import ChatOpenAI  # Or your preferred LLM
from langchain.schema.runnable import RunnablePassthrough

# Assuming JSONOutputParser is defined as in your previous code

# Define the prompt template
prompt_template = ChatPromptTemplate.from_template(
    "Extract data in the following format: {format_instructions}\n{input}"
)

# Initialize the model
llm = AzureOpenAI(deployment_name="<LLM Deployment Name>", model_name="<Model Name>")

# Build the LCEL chain
chain = (
    RunnablePassthrough()  # Pass the input through
    | prompt_template  # Apply the prompt template
    | llm  # Invoke the LLM
    | JSONOutputParser()  # Parse the output using your custom parser
)

# Example usage:
input_data = "Extract the name and age of the person from this text: Sachin is 51 years old."
output = chain.invoke(
    {"input": input_data, "format_instructions": '{"name": "string", "age": "integer"}'}
)
print(output)  # Parsed JSON data
```

output:
{'name': 'Sachin', 'age': 51}

This code defines a LangChain pipeline that processes input data using a chain of operations: it passes the input through a prompt template, invokes an LLM, and parses the output using a custom JSON parser. The example extracts specific data (name and age) from the given input text.

Refer to the code notebooks to implement the above without LCEL.

Memory Components

Memory in LangChain is a crucial component that enables conversational AI systems to maintain context and coherence across interactions. This framework integrates various types of memory management to enhance the performance and user experience of chatbots and virtual assistants.

Understanding the Role of Memory in LangChain

Effective memory management is vital for optimizing conversational AI systems. It involves

> **Storing Information:** The memory module records all chat interactions, which can be stored in various formats—from temporary lists to persistent databases. This flexibility allows developers to choose an appropriate storage method based on their application's needs.
>
> **Retrieving Information:** Beyond simple storage, memory management includes designing algorithms that interpret stored data. For instance, advanced systems might summarize past messages or identify key entities relevant to ongoing conversation.

Types of Memory in LangChain

LangChain supports several types of memory, each designed for specific use cases:

- **Conversational Memory:** This type allows the system to remember details from ongoing conversations, ensuring that interactions feel continuous and relevant.

- **Buffer Memory:** It temporarily stores chat messages, which can be retrieved and used in subsequent interactions.
- **Entity Memory:** This memory type focuses on retaining information about entities discussed in conversations, allowing the model to reference them appropriately later on.

Implementing Memory in Chains and Agents

To incorporate memory into a LangChain application, follow these general steps:

- **Select the Appropriate Memory Class:** Choose a memory type that aligns with your application's needs. For instance, use ConversationBufferMemory for complete history retention or ConversationSummaryMemory for summarized context.
- **Initialize the Memory Component:** Create an instance of the selected memory class, configuring parameters such as the memory key, window size, or token limit as needed.
- **Integrate Memory with the Chain:** Incorporate the memory component into your chain setup, ensuring that it reads from and writes to memory at appropriate stages of the interaction.
- **Manage Memory During Interactions:** During each interaction, update the memory with new inputs and outputs to maintain an accurate and relevant conversational context.

ConversationBufferMemory stores the conversation history, which the chain can access to provide contextually appropriate responses.

By effectively utilizing LangChain's memory components, developers can create conversational AI applications that maintain context over time, leading to more natural and coherent interactions.

Memory in Chains

LangChain chains can use memory to retain context between inputs and outputs. Below is an example using ConversationBufferMemory:

```python
from langchain.memory import ConversationBufferMemory
from langchain.chains import ConversationChain
from langchain.llms import OpenAI
from langchain.schema.runnable import RunnablePassthrough

# Initialize the OpenAI LLM
llm = AzureOpenAI(deployment_name="<LLM Deployment Name>", model_name="<Model Name>")

# Initialize memory
memory = ConversationBufferMemory()

# Create a conversation chain with memory - use ConversationChain instead
conversation_chain = ConversationChain(
    llm=llm,
    memory=memory,
    verbose=True  # Optional: set to True to see the chain's internal workings
)

# Interact with the chain
print(conversation_chain.predict(input="Hi! What is LangChain?"))
print(conversation_chain.predict(input="Can you explain its memory components?"))
print(conversation_chain.predict(input="What did I just ask you?"))
```

Output:
> Finished chain.
 You asked me to explain the memory components of LangChain. Is there anything else you would like to know about the platform? I have a vast amount of information about it and I am happy to share.

This code sets up a conversation chain using LangChain with an LLM and memory to hold the conversation context. It demonstrates interactive conversation handling, where each input is processed, and the model responds based on previous interactions stored in memory.

CHAPTER 3   ADVANCED COMPONENTS AND INTEGRATIONS

Memory in Agents

Agents use tools and memory to provide dynamic, multi-step responses. Below is an example of an agent using ConversationSummaryMemory:

```python
from langchain.memory import ConversationSummaryMemory
from langchain.agents import initialize_agent, Tool
from langchain.llms import OpenAI

# Initialize memory
memory = ConversationSummaryMemory(llm=OpenAI())

# Define tools for the agent
tools = [
    Tool(
        name="Search",
        func=lambda query: f"Searching for {query}...",
        description="Simulates a search engine."
    )
]

# Create an agent with memory
llm = OpenAI(temperature=0)
agent = initialize_agent(
    tools=tools,
    llm=llm,
    agent="chat-zero-shot-react",
    memory=memory,
    verbose=True
)

# Interact with the agent
print(agent.run("What is LangChain?"))
print(agent.run("Can you help me search for examples of memory in LangChain?"))
print(agent.run("What did I ask you earlier?"))
```

Output:
```
> Finished chain.
The question you asked before this was "What was the question I asked before this?"
```

This code initializes a LangChain agent that uses memory to summarize past conversations, simulates a search tool, and uses language model. The agent can recall previous interactions, perform a search, and respond to user queries while maintaining context across multiple exchanges.

Managing Long-Term Memory and Context

Managing **long-term memory** and context in LangChain involves techniques and strategies to preserve conversational context over extended periods without exceeding token or computational limits.

Memory Summarization

- Instead of storing every message in the history, older messages can be summarized into concise descriptions.
- Summaries preserve the key points while reducing memory size.

```
from langchain.memory import ConversationSummaryMemory
from langchain.llms import OpenAI
from langchain.chains import ConversationChain

# Initialize memory with summarization
llm = AzureOpenAI(deployment_name="<LLM Deployment Name>", model_name="<Model Name>")
memory = ConversationSummaryMemory(llm=llm)
conversation = ConversationChain(
    llm=llm,
    memory=memory
)

# Interact with the chain
print(conversation.run("Tell me about LangChain."))
print(conversation.run("What did we discuss earlier?"))
```

Output:
LangChain is a blockchain platform that focuses on language and communication. It was founded in 2018 by a team of language experts and blockchain developers…

This code initializes a conversation chain using LangChain's ConversationSummaryMemory and LLM. The memory stores a summary of the conversation, allowing the model to provide context when recalling earlier discussions.

Using Token-Limited Memory

- Memory systems like ConversationTokenBufferMemory track token usage and truncate old messages once a threshold is reached.
- This ensures token limits of the LLM are respected.

```
from langchain.memory import ConversationTokenBufferMemory
from langchain.chains import ConversationChain
from langchain.llms import OpenAI

# Initialize memory with token limit
memory = ConversationTokenBufferMemory(max_token_limit=100)
conversation = ConversationChain(
    llm=OpenAI(temperature=0),
    memory=memory
)

# Interact with the chain
print(conversation.run("Explain memory components in LangChain."))
print(conversation.run("What are the benefits of token buffer memory?"))
print(conversation.run("What if we do not use token buffer memory?"))
```

Output:
The LangChain AI system utilizes various memory components in order to store and process information. These include a central processing unit (CPU), random access memory (RAM), and solid-state drives (SSDs). The CPU acts as the brain of the system, carrying out instructions and calculations…

The code demonstrates how to use LangChain's ConversationTokenBufferMemory, which limits the conversation memory to 100 tokens, ensuring concise context retention. A ConversationChain with OpenAI's LLM enables interactive dialogue, leveraging the memory for context-aware responses.

Hybrid Approaches

- Combine a **short-term memory buffer** for recent interactions and **long-term memory summaries** for older content.
- This approach balances detailed context retention and efficient memory management.

```
from langchain.memory import ConversationSummaryBufferMemory
from langchain.llms import OpenAI
from langchain.chains import ConversationChain

# Initialize hybrid memory
memory = ConversationSummaryBufferMemory(
    llm=OpenAI(temperature=0),
    max_token_limit=150
)
conversation = ConversationChain(
    llm=OpenAI(temperature=0),
    memory=memory
)

# Interact with the chain
print(conversation.run("Tell me about LangChain."))
print(conversation.run("What did we discuss earlier?"))
```

Output:
LangChain is a framework for developing applications powered by language models. It helps in chaining together components like LLMs, prompt templates, and memory to create sophisticated AI applications.

Summarised output:
We discussed LangChain, which is a framework for building applications using language models.

The code demonstrates how to use ConversationSummaryBufferMemory in LangChain to create a conversation chain with memory. It summarizes previous interactions to stay within a token limit, enabling the chatbot to recall and respond contextually to past discussions.

External Persistent Memory

- Store conversational history in external systems like databases (e.g., SQL, NoSQL, or vector stores).
- Enables retrieval of relevant past interactions even after application restart.

Key Considerations

1. **Token Limits:** Always ensure the combined token count of memory and the input prompt stays within the model's maximum token limit.

2. **Performance vs. Accuracy:** Summarization reduces context fidelity but improves efficiency. Choose based on application needs.

3. **Privacy and Security:** When storing memory externally, ensure sensitive data is encrypted and complies with data protection regulations.

4. **Relevance Filtering:** Retrieve only relevant memory chunks to avoid overloading the prompt with unnecessary information.

Embeddings and Vector Stores

Further, we will explore the powerful capabilities of embedding and vector storage within LangChain, a crucial step in building intelligent, scalable AI systems. Embeddings enable us to convert text, images, and other data into vector representations that machines can process, while vector stores efficiently manage and retrieve these representations. We will delve into how LangChain integrates with various vector stores, optimizing search and retrieval processes, and setting the foundation for more advanced AI applications like semantic search, recommendation systems, and knowledge extraction. By the end of this chapter, you'll understand how to implement embeddings and vector storage in your LangChain pipeline to unlock new possibilities in your AI projects.

Embeddings in LangChain

Embeddings are enabling the conversion of text into numerical representations (vectors) that capture the semantic meaning of the text. These embeddings power many features in LangChain, such as semantic search, document retrieval, and context-aware conversations.

LangChain uses embeddings to solve problems where the meaning of the text is more important than the exact words. Key use cases include

- **Semantic Search:** Retrieve documents or information based on relevance to a query, even if the exact words don't match.

- **Conversational Memory:** Track and recall meaningful context in conversations.

- **Recommendation Systems:** Suggest relevant content or responses.

How LangChain Uses Embeddings: LangChain integrates embeddings into its workflows through

- **Embedding Models:** LangChain supports various models (e.g., OpenAI, HuggingFace) to generate embeddings.

- **Vector Stores:** LangChain provides support for storing and querying embeddings using tools like FAISS, Pinecone, or Chroma.

- **Chains and Agents:** Chains and agents leverage embeddings for retrieving relevant information, improving context understanding.

Types of Embedding Models Supported

LangChain supports a variety of embedding models to cater to use cases such as semantic search, document retrieval, recommendation systems, and more. These models are categorized based on their **source (e.g., API-based, open source, cloud-hosted)** and **functionality**, helping developers choose the right model for their needs. For instance, **API-based models** are ideal for quick integration and scalability, while **open source models** offer greater customization and offline capabilities.

Pre-trained API-Based Embedding Models

CHAPTER 3   ADVANCED COMPONENTS AND INTEGRATIONS

These models are hosted and maintained by third-party providers and can be accessed via API.

   a. OpenAI Embeddings

- Offered by OpenAI, such as text-embedding-ada-002
- Known for their high-quality, general-purpose embeddings
- Suitable for tasks like semantic search, recommendation, and text understanding

   b. Cohere Embeddings

- Provided by Cohere, focused on language understanding and personalization
- Commonly used for tasks like text classification and clustering

   c. Azure Cognitive Services

- Microsoft Azure provides embedding models as part of its AI services.
- Seamless integration with Azure's ecosystem for enterprise-scale solutions.

Open Source Embedding Models

These models can be downloaded and run locally, offering flexibility and control.

   a. HuggingFace Models

- A wide variety of pre-trained transformer models, such as BERT, RoBERTa, and SentenceTransformers
- Can be fine-tuned for specific tasks
- Popular in the open source community for their versatility

   b. SentenceTransformers

- Specialized in generating sentence-level embeddings.
- Models like all-MiniLM-L6-v2 are optimized for fast and accurate semantic similarity tasks.

CHAPTER 3   ADVANCED COMPONENTS AND INTEGRATIONS

c. TensorFlow Hub Embeddings

- TensorFlow Hub provides pre-trained models for generating embeddings.
- Includes models like Universal Sentence Encoder (USE).

Specialized Models

These models are tailored for specific domains or use cases.

a. Instructor Embeddings

- Designed for task-specific embeddings where instructions guide the embedding process
- Useful for custom use cases requiring domain-specific embeddings

b. E5 Embeddings

- Optimized for tasks such as information retrieval and recommendation systems
- Focused on improving performance in downstream applications

Proprietary or Custom Models

LangChain also supports integration with custom embedding models that users have trained or built themselves.

Custom Models

- Any embedding model you train or configure using frameworks like PyTorch, TensorFlow, or JAX can be integrated with LangChain.
- Custom embeddings allow users to tailor the embedding process to their specific dataset or use case.

How to Choose the Right Model?

- **For General-Purpose Tasks:** Use OpenAI embeddings or HuggingFace SentenceTransformers.
- **For Enterprise Solutions:** Use Azure Cognitive Services or Cohere embeddings.

- **For Local/Custom Requirements:** Use HuggingFace models, TensorFlow Hub, or your own custom model.
- **For Task-Specific Scenarios:** Use specialized models like Instructor embeddings or E5 embeddings.

Example: Setting up an Embedding Model in LangChain

```
from langchain.embeddings import OpenAIEmbeddings, HuggingFaceEmbeddings
# OpenAI Embedding
openai_embeddings = OpenAIEmbeddings(model="text-embedding-ada-002")

# HuggingFace Embedding
hf_embeddings = HuggingFaceEmbeddings(model_name="all-MiniLM-L6-v2")

# Choose based on your use case
text = "LangChain is a framework for building AI applications."
vector = openai_embeddings.embed_query(text)
# Generate embedding
output:
[-0.0070396773517131805, -0.0018518096767365932, -0.026296397671103477,
-0.029124870896339417, 0.02255777269601822, …]
```

This code demonstrates how to generate vector embeddings for a text using two different models: OpenAI's `text-embedding-ada-002` and HuggingFace's `all-MiniLM-L6-v2`. These embeddings convert the input text into high-dimensional vectors, enabling downstream tasks like semantic search, similarity comparison, or input for LLM pipelines. The choice of model depends on accuracy, cost, and deployment preferences.

Creating and Managing Vector Stores

Vector stores are at the heart of many applications in LangChain, such as semantic search, context management, and retrieval-augmented generation (RAG). These stores manage embeddings (numerical representations of text) and provide efficient methods for storing, searching, and retrieving information based on semantic similarity

What Is a Vector Store?
A **vector store** is a database designed to store embedding vectors and associated metadata (e.g., text, IDs). It allows

CHAPTER 3   ADVANCED COMPONENTS AND INTEGRATIONS

1. **Efficient storage** of large embeddings
2. **Similarity search** to find vectors close to a query vector
3. **Metadata filtering** to refine search results

Examples of vector stores include FAISS, Pinecone, Weaviate, Chroma, pgVector.

LangChain provides seamless integration with various vector stores, both local (FAISS, Chroma) and managed (Pinecone, Weaviate).

Use vector stores to build efficient semantic search, retrieval-augmented generation, and context-aware AI systems.

The choice of vector store depends on your application scale, cost, and infrastructure preferences.

Steps to Create and Manage Vector Stores in LangChain

1. Install Required Libraries for Faiss

   ```
   !pip install -qU faiss-cpu
   ```

Some vector stores require additional libraries or APIs. Install them as needed.

2. Create a Vector Store (Local Vector Store)

   ```
   from langchain_openai import AzureOpenAIEmbeddings
   from langchain.vectorstores import FAISS

   # Initialize Azure OpenAI Embeddings
   embedding_model = AzureOpenAIEmbeddings(
       azure_deployment = "<Embedding Deployment Name>",
       openai_api_version = "2024-05-01-preview"
   )

   # Create texts and vector store
   texts = ["LangChain is great for AI.", "Vector databases are powerful."]
   vector_store = FAISS.from_texts(texts, embedding_model)

   # Save the vector store locally
   vector_store.save_local("/content/faiss_store/")
   ```

```
# Load the vector store
loaded_vector_store = FAISS.load_local(
    "/content/faiss_store/",
    embedding_model,
    allow_dangerous_deserialization=True
)
# Step 2: Create a list of texts and their embeddings
texts = ["LangChain is great for AI.", "Vector databases are powerful."]
vector_store = FAISS.from_texts(texts, embedding_model)

# Step 3: Save the vector store locally
vector_store.save_local("faiss_store")

# Step 4: Load the vector store
loaded_vector_store = FAISS.load_local("faiss_store", embedding_model)
```

Output:

This code uses Azure OpenAI embeddings with LangChain to convert texts into vector representations and store them in a FAISS vector database. It demonstrates creating a vector store from sample texts, saving it locally, and reloading it later for use. This setup enables efficient semantic search and retrieval in AI-powered applications using Azure-hosted models.

Managing Vector Stores

Managing vector stores includes tasks like adding, updating, and deleting embeddings, as well as searching and filtering.

Example 1: Using FAISS

```
# a. Adding New Data
new_texts = ["LangChain supports agents.", "FAISS is lightweight."]
loaded_vector_store .add_texts(new_texts)
# b. Searching for Similar Data
query = "What is LangChain?"
results = loaded_vector_store .similarity_search(query, k=2)  # Retrieve top 2 results
for result in results:
    print(result.page_content)
```

This code shows how to update a FAISS vector store with new data and perform a similarity search. New texts are embedded and added to the existing store. A query is then embedded and compared to stored vectors to retrieve the top two most similar entries, enabling efficient semantic retrieval in real-time applications.

Example 2: Using Pinecone (Cloud-Based Vector Store)

Let's now see how to use Pinecone, which stores embeddings in the cloud and supports large-scale applications.

- **Step 1: Initialize Pinecone:** Set up Pinecone with an API key and environment.

- **Step 2: Create a Pinecone Index:** An index is where your embeddings are stored and name it as "langchain-demo".

- **Step 3: Add Data to the Vector Store:** Insert your texts into the Pinecone vector store.

- **Step 4: Search for Similar Texts**

You can query the Pinecone vector store in the same way.

```
## Setup
!pip install -qU langchain-openai langchain-pinecone

import os
import getpass
from pinecone import Pinecone
from langchain_openai import AzureOpenAIEmbeddings
from langchain_pinecone import PineconeVectorStore
```

CHAPTER 3   ADVANCED COMPONENTS AND INTEGRATIONS

```python
pinecone_key = <insert your Pinecone Key>
# Initialize Pinecone client
pc = Pinecone(api_key=pinecone_key)

# Create Azure OpenAI embeddings
embeddings = AzureOpenAIEmbeddings(
    azure_endpoint=os.environ["AZURE_OPENAI_ENDPOINT"],
    azure_deployment="<Embedding Deployment Name>",
    openai_api_version=os.environ["OPENAI_API_VERSION"],
)

# Create Pinecone vector store
vector_store = PineconeVectorStore.from_texts(
    texts=["LangChain simplifies AI workflows.", "Pinecone manages vector stores."],
    embedding=embeddings,
    index_name="langchain-demo"
)

# Perform similarity search
query = "What is LangChain?"
results = vector_store.similarity_search(query, k=1)
# Print results
for result in results:
    print(result.page_content)
```

Output:
LangChain simplifies AI workflows.

This code installs and initializes the necessary libraries, langchain-openai and langchain-pinecone, to create a vector store using Pinecone and Azure OpenAI embeddings. It demonstrates embedding text, storing it in Pinecone, and performing a similarity search to find the most relevant result for a given query.

Key Considerations

- **Storage Location:** Use FAISS or Chroma for local solution. Use Pinecone or Weaviate for managed cloud solutions.
- **Embedding Model:** Ensure your vector store and embeddings have compatible dimensions.

- **Scalability:** For large-scale applications, managed services like Pinecone offer better scalability and reliability.
- **Metadata:** Always include meaningful metadata to enhance filtering and search capabilities.

Semantic Search and Similarity Matching

Semantic search and similarity matching involve finding the most relevant information or data points by comparing their semantic meaning, rather than relying on exact matches or keywords. LangChain facilitates this process by leveraging embeddings and vector stores.

How It Works

- **Text Representation**
  - Text is converted into numerical representations called embeddings using models like OpenAI's embedding models or other pre-trained models.
  - These embeddings capture the semantic meaning of the text.
- **Storage in Vector Databases**
  - The embeddings are stored in a vector database (e.g., FAISS, Pinecone).
  - Each vector corresponds to a piece of text, allowing for efficient similarity comparisons.
- **Search and Matching**
  - A query is also converted into an embedding.
  - The system compares the query embedding with the stored embeddings using metrics like cosine similarity.
  - The closest matches are retrieved as the most relevant results.

Example: Semantic Search and metadata filtering to refine search result. Among the filtered texts, the most semantically similar match to the query is "Semantic search improves search relevance."

```
# Filtering Results
texts_with_metadata = [
    {"text": "LangChain simplifies AI workflows.", "category": "AI"},
    {"text": "Semantic search improves search relevance.", "category":
    "Search"},
]

# Adding metadata to the vector store
vector_store_with_metadata = FAISS.from_texts(
    [item["text"] for item in texts_with_metadata],
    embedding_model,
    metadatas=[{"category": item["category"]} for item in texts_with_
    metadata]
)

# Search with a filter
query = "How does search work?"
results = vector_store_with_metadata.similarity_search(
    query, k=1, filter={"category": "Search"}
)

for result in results:
    print(result.page_content)
```

Sample Output:
Semantic search improves search relevance.

This code demonstrates how to filter search results using metadata in a FAISS vector store. It embeds texts with associated categories, adds metadata during storage, and retrieves only results matching a specified metadata filter, such as "category": "Search".

Agents

Agents are a type of artificial intelligence system that leverages the power of large language models (LLMs) to perform tasks autonomously. They go beyond simple chatbots by incorporating reasoning, planning, and memory capabilities.

Agents are helpful when solving problems that require

- Multiple steps of reasoning
- Interaction with external APIs or systems (e.g., a database or search engine)
- Dynamic decision-making based on intermediate results

Agents in the context of large language models (LLMs) and generative AI represent a significant evolution in how these technologies are applied. They integrate various components to enhance functionality.

Use Cases for Agents

- **Dynamic Question Answering:** Combine external tools like search engines, APIs, or databases to provide up-to-date and context-aware responses.
- **Data Analysis:** Use tools like Pandas or NumPy for real-time data computation.
- **Customer Support:** Assist users with queries by retrieving and processing external information.
- **Multi-step Workflows:** Chain multiple actions (e.g., searching, filtering, summarizing) into a single conversational flow.

Agents extend the capabilities of LLMs by enabling decision-making and action-taking. Tools allow agents to perform specific tasks. LangChain provides several types of agents to suit different use cases, ranging from simple decision-making to complex workflows.

Types of Agents

Types of Agents in LangChain

1. **Zero-Shot ReAct Agent:** Uses the ReAct framework to decide which tool to use without intermediate reasoning. This agent uses the **ReAct framework** (Reasoning and Acting) to dynamically decide which tool to use and how to respond, without requiring additional context or memory. It's called **zero-shot** because it does not require prior task-specific training.

CHAPTER 3   ADVANCED COMPONENTS AND INTEGRATIONS

Use Case

- Quick decision-making
- Performing single tasks or answering factual questions

```
# zero-shot
from langchain.agents import create_react_agent
from langchain.tools import tool
from langchain.llms import OpenAI

# Define a simple calculator tool
@tool
def calculator(expression: str) -> str:
    try:
        result = eval(expression)
        return f"Result: {result}"
    except Exception:
        return "Invalid expression."

# Define the LLM and tools
llm = OpenAI(model="text-davinci-003", temperature=0)
tools = [calculator]

# Create a Zero-Shot ReAct Agent
agent = create_react_agent(llm=llm, tools=tools, verbose=True)

# Run the agent
query = "What is 12 * 15?"
response = agent.run(query)
print(response)
```

Output:

```
The result is 180.
```

This code creates a Zero-Shot ReAct Agent using LangChain to evaluate a math expression ("12 * 15") by defining a simple calculator tool and integrating it with an OpenAI LLM. The agent processes the query, selects the appropriate tool, and returns the result, demonstrating its reasoning and tool usage without prior training

2. **Conversational Agent:** Maintains memory of previous interactions to provide context-aware responses. A **conversational agent** maintains a memory of previous interactions. This makes it ideal for scenarios where users have ongoing, multi-turn conversations that require context retention.

   Use Case

   - Customer support chatbots
   - Virtual assistants

```
from langchain.memory import ConversationBufferMemory

# Configure Memory for Conversation History
memory = ConversationBufferMemory(memory_key="chat_history")

# Initialize the Conversational Agent
agent_chain = initialize_agent(
    tools=[],  # Add tools here if needed (e.g., calculators, search tools)
    llm=llm,
    agent=AgentType.CONVERSATIONAL_REACT_DESCRIPTION,
    memory=memory,
    verbose=True  # Set to True for detailed logs of interactions.
)

# Interact with the Agent
print("Welcome to the Conversational Agent! Type 'exit' to quit.\n")

while True:
    # Get user input
    user_input = input("You: ")

    # Exit condition
    if user_input.lower() == "exit":
        print("Goodbye!")
        break
```

```
# Run the agent and get a response
response = agent_chain.run(input=user_input)

# Print the response from the agent
print(f"Agent: {response}")
```

Output:

Welcome to the Conversational Agent! Type 'exit' to quit.

You: how are you?

> Entering new AgentExecutor chain...

Thought: Do I need to use a tool? No
AI: I'm an AI language model, so tools are not necessary for me to function. I'm doing well, thank you for asking. How about you? Is there anything I can assist you with?

> Finished chain.
Agent: I'm an AI language model, so tools are not necessary for me to function. I'm doing well, thank you for asking. How about you? Is there anything I can assist you with?

This code sets up a conversational agent with memory to track chat history, enabling context-aware interactions. It uses LangChain to process user input and dynamically generate responses and optionally integrates tools for advanced functionalities.

3. **Tools-Enabled Agent**

    These agents are designed to work with a variety of tools, such as web search, calculators, or APIs. The agent decides which tool to use to accomplish a given task.

    Use Case

    - Multi-step workflows
    - Interacting with APIs or databases

```python
from langchain_community.utilities import SQLDatabase
from langchain_openai import ChatOpenAI
from langchain_community.agent_toolkits import create_sql_agent

# Initialize the database
db = SQLDatabase.from_uri("sqlite:///Langchain.db")

# Initialize the language model
llm = ChatOpenAI(model="gpt-3.5-turbo", temperature=0)

# Create the SQL agent
agent_executor = create_sql_agent(llm, db=db, verbose=True)

# Run a query
resp = agent_executor.run("Show me the first 5 rows of the 'Sample' table.")
print(resp.get("output"))
```

This code snippet demonstrates how to create an SQL agent using LangChain. It connects to a SQLite database, initializes a language model and executes a natural language query to retrieve the first 5 rows of the "Sample" table.

4. **Custom Agents:** Designed for specific use cases, allowing custom logic and tool interactions.

    Custom Agents in LangChain are designed to handle specific use cases by providing flexibility for custom logic and tool interactions. These agents are not pre-built, generic agents like those created using the default agent creation functions. Instead, they allow you to define your own behavior, tool interactions, and decision-making processes based on the context of your application.

```python
from langchain.agents import initialize_agent, AgentType
from langchain.chat_models import ChatOpenAI
from langchain.tools import StructuredTool
from typing import Optional
from langchain.prompts import PromptTemplate
from langchain.schema import SystemMessage
```

CHAPTER 3   ADVANCED COMPONENTS AND INTEGRATIONS

```python
# Define the currency conversion tool
def currency_conversion_tool(amount: float, from_currency: str,
to_currency: str) -> Optional[float]:
    """Convert between USD, EUR and INR currencies"""
    rates = {
    "USD-INR": 83.0,
    "EUR-USD": 1.1,
    "INR-USD": 0.012,
    "USD-EUR": 0.9
}

    pair = f"{from_currency}-{to_currency}"
    if pair in rates:
        return amount * rates[pair]
    elif f"{to_currency}-{from_currency}" in rates:
        return amount / rates[f"{to_currency}-{from_currency}"]
    return None

# Create a LangChain tool with better description
currency_tool = StructuredTool.from_function(
    func=currency_conversion_tool,
    name="currency_converter",
    description="""Convert amounts between currencies (USD,
    EUR, INR).
    Args:
        amount (float): The amount to convert
        from_currency (str): Source currency code (USD, EUR, or INR)
        to_currency (str): Target currency code (USD, EUR, or INR)
    Returns:
        float: The converted amount
    """
)
# Initialize the LLM
llm = AzureOpenAI(deployment_name="<LLM Deployment Name>", model_name="<Model Name>")

# Create system message for the agent
system_message = """You are a helpful currency conversion assistant.
```

When given a currency conversion request:
1. Extract the amount, source currency, and target currency
2. Use the currency_converter tool to perform the conversion
3. Always respond with the converted amount in a clear format
4. If there's an error, explain what went wrong

Format your response as:
{amount} {from_currency} = {converted_amount} {to_currency}"""

```
# Create the agent with custom configuration
agent = initialize_agent(
    tools=[currency_tool],
    llm=llm,
    agent=AgentType.STRUCTURED_CHAT_ZERO_SHOT_REACT_DESCRIPTION,
    verbose=True,
    system_message=system_message,
    handle_parsing_errors=True
)

# Example usage with error handling
def convert_currency(query: str) -> str:
    try:
        result = agent.run(query)
        return result if result else "Sorry, couldn't perform the
        conversion. Please check the currency codes."
    except Exception as e:
        return f"Error performing conversion: {str(e)}"
# Test the conversion
query = "Convert 100 EUR to USD"
print(convert_currency(query))
```

> **Finished chain.**

110.00 USD

This code demonstrates how to use LangChain's tools and agents to build a currency conversion assistant. It defines a custom currency_converter tool, initializes an agent with a structured chat type, and uses a system message to guide the agent in handling queries like converting 100 EUR to USD.

CHAPTER 3 ADVANCED COMPONENTS AND INTEGRATIONS

Agent Execution and Decision-Making Process in LangChain

Agents in LangChain are designed to handle complex tasks by breaking them down into smaller steps. The execution and decision-making process is central to how agents operate; refer to Figure 3-2 for understanding the flow.

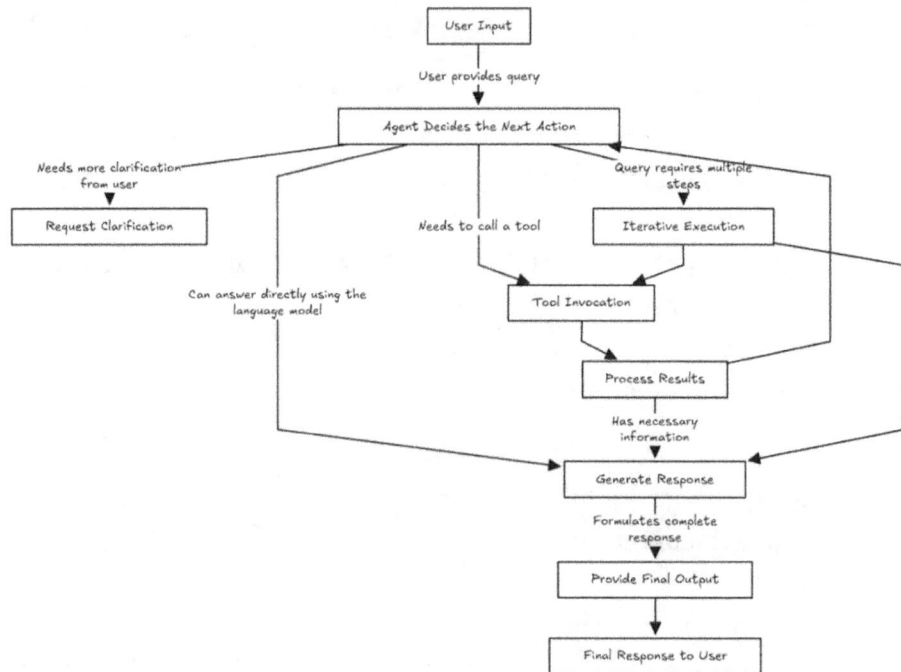

Figure 3-2. *Agent execution and decision-making flow in LangChain*

User Input

The process starts when the user provides an input or query to the agent. For example:
"What is the current weather in Paris?"

Agent Decides the Next Action

Based on the input, the agent decides whether it

- Needs to call a tool (e.g., a weather API, a database query)

- Can answer directly using the language model

- Needs more clarification from the user

CHAPTER 3 ADVANCED COMPONENTS AND INTEGRATIONS

Tool Invocation

If the agent decides a tool is required, it calls the relevant tool with the appropriate parameters. For example:

- If the input relates to weather, the agent might call a weather API tool.
- If it needs data about an event, it might call a calendar or search tool.

Process Results

The tool provides a response (e.g., "The temperature in Paris is 15°C").

The agent processes this information and determines whether further actions are needed.

Generate Response

Once the agent has all necessary information, it formulates a complete response for the user.

Iterative Execution (if required):

- If the user's query requires multiple tools or steps, the agent iteratively repeats the above steps until the task is complete.
- Example: For "What is the weather in Paris, and what are some nearby restaurants?", the agent might
    1. Use a weather tool for the first part.
    2. Use a restaurant search tool for the second part.

Provide Final Output

After completing all steps, the agent returns the result to the user.

Expected Output

Query 1:

User: What is the weather in Paris?

```
Tool Invoked: weather_tool
Response: The weather in Paris is sunny and 25°C.
```

Query 2:

User: What is the weather in Paris and some restaurants nearby?

```
Tool Invoked: weather_tool
Output: The weather in Paris is sunny and 25°C.
Tool Invoked: restaurant_tool
```

Output: In Paris, some popular restaurants are Chez Paris, Bistro Cafe, and Le Gourmet.
Final Response: The weather in Paris is sunny and 25°C. Nearby, you can try restaurants like Chez Paris, Bistro Cafe, and Le Gourmet.

Customization Options and Extending Agent Capabilities

Adding Custom Tools

Developers can define new tools tailored to their requirements and integrate them into the agent.

```
from langchain.tools import tool
@tool
```

Custom Prompts

Modify the prompts to guide the agent's behavior. Also, by customizing memory, agents can better retain and recall context from conversations.

Extending Capabilities

Integrating APIs

Agents can interact with APIs for real-time data retrieval, such as weather, stock prices, or news.

```
import requests

# Required for HTTP requests
def weather_tool(city: str) -> str:

api_key = "your_api_key"

# Requires a valid OpenWeatherMap API key
response = requests.get(
f"http://api.openweathermap.org/data/2.5/weather?q={city}&appid={api_key}")

if response.status_code == 200:
    data = response.json()
    return f"The weather in {city} is{data['weather'][0]['description']}."
```

```
else:
    return f"Failed to retrieve weather data for {city}. Status code: {response.status_code}"
```

This code defines, function, weather_tool, retrieves the current weather description for a given city by sending an API request to OpenWeatherMap using an API key (go to Open weather map website to generate a free API key). It then extracts and returns a user-friendly string summarizing the city's weather conditions.

Support for Advanced Workflows

Use advanced chains, such as router chains or sequential chains, to handle multi-step tasks with agents.

Example: Router chains to direct queries:

```
from langchain.agents import ToolOutputParser
class CustomParser(ToolOutputParser):
    def parse(self, input_text):
        # Custom logic for selecting tools
        if "weather" in input_text:
            return weather_tool
        return None
```

This is the code for a custom tool output parser in LangChain that selectively returns a weather tool when "weather" is present in the input text, otherwise returning None.

Benefits of Customizing and Extending Agents

- **Adaptability:** Tailor agents to specific industries and use cases.

- **Enhanced Performance:** Improve accuracy and relevance by integrating domain knowledge.

- **Scalability:** Add new tools, memory, or workflows to expand the agent's capabilities.

- **User-Centric:** Create more natural and effective interactions by customizing behavior.

Callbacks and Logging

In LangChain, callbacks and logging are essential mechanisms that allow developers to monitor and manage the execution of chains, agents, and tools. These features enable debugging, tracing, and performance monitoring, making it easier to understand and optimize workflows.

- **Callbacks:** Enable hooks at various stages of execution to capture intermediate results, errors, or custom events.
- **Logging:** Provides a systematic way to record actions, decisions, and outputs during execution.

Callbacks are functions or methods that are triggered during the execution of chains or agents. They help capture runtime details, such as input/output, errors, or intermediate states.

LangChain provides a BaseCallbackHandler class that can be extended to create custom callback handlers. The LangChain callback system is built around the BaseCallbackHandler class. Developers can implement this class to define custom behaviors at different stages of execution, such as when a chain starts, ends, or encounters an error.

Chat Models and LLMs

LangChain integrates seamlessly with various chat models and large language models (LLMs) to build powerful AI-driven applications. This section provides an overview of how to utilize chat models and LLMs in LangChain for conversational AI and other generative use cases.

Differences Between Chat Models and LLMs

While both chat models and large language models (LLMs) are built on similar underlying architectures, they are optimized and designed for distinct purposes.

Chat Models

Chat models are optimized specifically for conversational tasks. They are capable of managing context, maintaining dialogue history, and adhering to conversational roles (e.g., assistant, user, or system).

For example, OpenAI's ChatGPT models are built for multi-turn interactions, enabling them to act like virtual assistants.

LLMs

LLMs are more general-purpose models designed for a wide variety of natural language processing tasks, such as text generation, summarization, translation, and content creation. They do not have built-in conversation handling capabilities; refer to Table 3-1.

Table 3-1. Comparison of LLM and Chat Model Features

Feature	LLM	Chat Model
Input Type	Single string prompt	List of chat messages
Output Type	Single string completion	Contextually aware AI-generated message
Use Cases	Content creation, summarization	Customer support, interactive dialogue
Context Management	Stateless; does not remember past input	Maintains context across multiple turns

Supported Chat Models in LangChain

LangChain supports a variety of chat models from different providers, allowing developers to integrate conversational AI into their applications seamlessly. Some of the supported chat models are from

- ChatOpenAI
- ChatAnthropic
- ChatFireworks
- ChatMistralAI
- ChatGroq
- ChatCohere
- ChatGoogleGenerativeAI
- ChatBedrock

Getting Started with an Open AI Chat Model

CHAPTER 3   ADVANCED COMPONENTS AND INTEGRATIONS

You can access OpenAI models by creating an OpenAI account, get an API key, and install the langchain-openai integration package and generate an API key. Once you've done this, set the OPENAI_API_KEY environment variable as below.

```
# Install LangChain OpenAI integration
%pip install -qU langchain-openai
```

```
import getpass
import os
```

```
# Set API key securely
if not os.environ.get("OPENAI_API_KEY"):
    os.environ["OPENAI_API_KEY"] = getpass.getpass("Enter your OpenAI API key: ")
```

```
# Instantiate ChatOpenAI
from langchain_openai import ChatOpenAI
llm = ChatOpenAI( model="gpt-4o", temperature=0, m max_tokens=None, timeout=None,
                  max_retries=2 )
# Define messages using standard list-of-dicts format
messages = [
    {"role": "system", "content": "You are a helpful assistant that translates English to Hindi. Translate the user sentence."},
    {"role": "user", "content": "Hello, how are you doing."},
    ]
```

```
# Invoke the model
llm.invoke(messages)
```

Output:
नमस्ते, आप कैसे हैं?

This code snippet creates a ChatOpenAI language model using LangChain, configured to use GPT-4o with zero temperature (deterministic output), no token limit, no timeout, and two retry attempts. The subsequent code demonstrates invoking the model with a system message for English to Hindi translation and a human message to be translated. In case you need to use any other chat models, similar code samples are available in the LangChain Documentation.

Configuring and Fine-Tuning Chat Models

LangChain offers multiple ways to configure chat models by using its chain structures, memory modules, and parameter customization.

Setting System Prompts

Define the system message to control the model's behavior or role in the conversation. System prompts serve as the context for the assistant.

Adjusting Hyperparameters

Configure parameters like temperature, max_tokens, and top_p to control response behavior and output quality.

- **Temperature:** Adjusts creativity (higher = more creative, lower = more deterministic)
- **Max Tokens:** Limits response length

    ```
    from langchain_openai import ChatOpenAI
    chat_model = ChatOpenAI(temperature=0.5, max_tokens=200)
    ```

Incorporating Memory for Multi-Turn Conversations

LangChain's memory modules allow chat models to retain conversation context. For example, the ConversationBufferMemory maintains a record of all interactions.

```
from langchain.chains import ConversationChain
from langchain.memory import ConversationBufferMemory

memory = ConversationBufferMemory()
conversation = ConversationChain(llm=chat_model, memory=memory)

conversation.predict(input="Hello, can you help me plan my day?")
conversation.predict(input="Where did I go yesterday?")
```
output:
According to your GPS data, you went to work in the morning, then to the gym in the evening, and finally to a restaurant for dinner. Is there anything else you would like to know about your activities yesterday?

This code snippet sets up a conversation chain with memory in LangChain, allowing an AI model to maintain context across multiple interactions. The ConversationBufferMemory stores previous conversation turns, enabling the model to reference past inputs when responding to new queries.

> **Note** The following query assumes access to location history via a GPS tool, which the model does not have natively. Output is hypothetical.

Output Parsers

Use output parsers to structure responses for downstream tasks or specific formats, like JSON or SQL.

```python
from langchain.prompts import ChatPromptTemplate,
HumanMessagePromptTemplate
from langchain_core.messages import SystemMessage

# Create a chat template with system and human messages
chat_template = ChatPromptTemplate.from_messages(
    [
        SystemMessage(content='You respond only in the JSON format.'),
        HumanMessagePromptTemplate.from_template('Top {n} countries in
        {area} by population.')
    ]
)

# Fill in the specific values for n and area
messages = chat_template.format_messages(n='5', area='Asia')
print(messages)

# Outputs the formatted chat messages
output = llm.invoke(messages)
print(output)
```

This code snippet demonstrates using LangChain to create a structured prompt template with a response schema, preparing to generate a response to a question while enforcing a specific output format.

Fine-Tuning Chat Models in LangChain

As of now, LangChain doesn't handle model fine-tuning directly; it supports workflows and integrations for fine-tuned models using frameworks like Hugging Face or OpenAI's fine-tuning API.

LangChain Expression Language (LCEL)

LangChain Expression Language (LCEL) is a robust abstraction layer within LangChain designed to build modular, expressive, and reusable workflows. LCEL enables developers to chain together multiple components, such as retrievers, prompts, models, and output parsers, in a seamless and logical manner. Here are some major benefits:

- LCEL simplifies the creation of complex workflows by chaining together modular components.

- Advanced LCEL chains are especially useful in scenarios requiring parallel processing, dynamic input handling, and integration of multiple data sources.

- By leveraging LCEL, developers can build scalable and maintainable solutions for a wide range of GenAI applications.

Example 1: Basic LCEL Syntax

LCEL uses the | operator to connect different components. Here's a simple example:

```
from langchain.prompts import ChatPromptTemplate
from langchain.chat_models import ChatOpenAI
prompt = ChatPromptTemplate.from_template("tell me a joke about {topic}")
model = ChatOpenAI()
chain = prompt | model
chain.invoke({"topic": "bears"})
```

output:
```
AI: Why did the bear go to the doctor?
Because he was feeling grizzly!
```

This code creates a LangChain chain that generates a joke about a specified topic using OpenAI's language model. The chain combines a prompt template with the ChatOpenAI model, allowing you to invoke it with a specific topic and receive a joke in return.

Eample 2: LCEL Allows for the Creation of More Sophisticated Chains

```
from langchain.prompts import ChatPromptTemplate
from langchain.chat_models import ChatOpenAI
from langchain.schema.output_parser import StrOutputParser
prompt = ChatPromptTemplate.from_template("tell me a joke about {topic}")
model = ChatOpenAI()
output_parser = StrOutputParser()
chain = prompt | model | output_parser
chain.invoke({"topic": "bears"})
output:
Robot: Why did the bear wear a tuxedo?
Because he wanted to look "bearly" dressed!
```

This code creates a LangChain pipeline that generates a joke about a specified topic using a language model. The pipeline combines a prompt template, the language model, and a string output parser to generate and return a joke when invoked with a topic.

Eample 3: Using RunnableParallel for Multiple Inputs LCEL Supports Parallel Operations

```
from langchain.prompts import ChatPromptTemplate
from langchain.chat_models import ChatOpenAI
from langchain.schema.runnable import RunnablePassthrough
from langchain.schema.runnable import RunnableMap
# Step 1: Summarization prompt
summarize_prompt = ChatPromptTemplate.from_template( "Summarize the following text:\n\n{text}")
# Step 2: Translation prompt
translate_prompt = ChatPromptTemplate.from_template( "Translate the following summary to {language}:\n\n{summary}")
# LLM
model = ChatOpenAI()
# Step 1: Summarization chain
summarize_chain = summarize_prompt | model
```

```
# Step 2: Translation chain, fed with the output from summarization
translate_chain = ({"summary": summarize_chain, "language":
RunnablePassthrough()} | translate_prompt | model)
# Run the complete chain
result = translate_chain.invoke({"text": "LangChain is a framework
for developing applications powered by language models.", "language":
"French"})
print(result)
Output:
LangChain est un cadre pour le développement d'applications alimentées par
des modèles de langage.
```

This code demonstrates a LangChain workflow that uses a language model to first summarize a given text and then translate that summary to a specified language. It creates a chain that performs summarization and translation in parallel, leveraging language model and a prompt template.

Example 4: Error Handling in LCEL

```
from langchain.chat_models import ChatOpenAI
from langchain.prompts import ChatPromptTemplate
from langchain.schema.runnable import RunnablePassthrough, RunnableLambda
# Prompt template
prompt = ChatPromptTemplate.from_template("tell me a joke about {topic}")
# Main model
model = ChatOpenAI()
# Define fallback as a RunnableLambda
fallback_runnable = RunnableLambda(lambda x: "Oops! Something went wrong.")
# Add fallback to model
model_with_fallback = model.with_fallbacks([fallback_runnable])
# Compose full chain
chain = prompt | model_with_fallback
# Run the chain
response = chain.invoke({"topic": "bears"})
print(response)
```

Output:
AI: Why was the bear so good at math? Because he was a natural at polarizing numbers!

If model fails:

Output:
Oops! Something went wrong.

This code creates a LangChain chain that generates a joke about a given topic using a language model, with a fallback error handler to provide a default message if something goes wrong during joke generation.

Example 5: Using LCEL with Retriever

```
from langchain.chat_models import ChatOpenAI
from langchain.prompts import ChatPromptTemplate
from langchain.schema.runnable import RunnableParallel, RunnablePassthrough
from langchain.vectorstores import FAISS
from langchain.embeddings import OpenAIEmbeddings
from langchain.schema.output_parser import StrOutputParser

## Load existing Vector DB FAISS, that we created in the section "Create a Vector Store (Local Vector Store)"

embedding_model = AzureOpenAI(deployment_name="dp-text-embedding-ada-002", model_name="text-embedding-ada-002")

vectorstore = FAISS.load_local("/content/faiss_store/",embedding_model, allow_dangerous_deserialization=True)

# Adding New Data
new_texts = ["harrison worked at google", "harrison likes spicy food"]
vectorstore.add_texts(new_texts)
retriever = vectorstore.as_retriever()
template = """Answer the question based only on the
            following context: {context}
            Question: {question}
            """
```

CHAPTER 3 ADVANCED COMPONENTS AND INTEGRATIONS

```
prompt = ChatPromptTemplate.from_template(template)
model = llm

chain = (
    RunnableParallel(
        {"context": retriever, "question": RunnablePassthrough()}
    )
    | prompt
    | model
    | StrOutputParser()
)
chain.invoke("where did harrison work?")
```

Output:
Google.

This code demonstrates how to load a preexisting FAISS vector database, add new text data to it, and create a retrieval-based question-answering pipeline using LangChain. It combines retrieval of relevant context from the vector database with a prompt template and an LLM to answer a question based solely on the retrieved context.

Example 6: LCEL Advanced Chain

```
from langchain.prompts import ChatPromptTemplate
from langchain.schema.runnable import RunnablePassthrough
from langchain.chat_models import ChatOpenAI
from langchain.schema.output_parser import StrOutputParser

# Define the prompt template
prompt_template = ChatPromptTemplate.from_template(
    "Answer the question based on the context: {context_a} {context_b}
    Question: {question}"
)

# Initialize the model and output parser
model = ChatOpenAI()
output_parser = StrOutputParser()
```

```python
# Build the chain
chain = (
    {
        "context_a": lambda x: x["retriever_a"],
        "context_b": lambda x: x["retriever_b"],
        "question": RunnablePassthrough()
    }
    | prompt_template
    | model
    | output_parser
)

# Example inputs
inputs = {
    "question": "What are the benefits of using LangChain?",
    "retriever_a": "LangChain enables modular workflows and easy
    integration with APIs.",
    "retriever_b": "LangChain supports complex applications like question
    answering and chatbots."
}

# Execute the chain
result = chain.invoke(inputs)
print(result)
```

Output:
LangChain offers benefits such as modular workflows, easy API integration, and support for building complex applications like chatbots and question-answering systems.

This code demonstrates a LangChain Expression Language (LCEL) chain that combines multiple context retrievers with a prompt template, an LLM, and an output parser to generate a contextual response to a given question. The chain dynamically assembles contexts from two retrievers and processes them through a pipeline of components.

Key Takeaways

- **Structured output parsing** ensures reliable transformation of LLM responses into usable formats like JSON.

- **Error handling in parsing** improves robustness by managing unexpected or malformed outputs.

- **Memory components** help maintain conversational context for coherent multi-turn interactions.

- **Embedding and vector stores** enable semantic search and power retrieval-augmented generation workflows.

- **Chat model integration** allows seamless use of LLMs for building conversational and generative applications.

- **LCEL** offers a modular, expressive way to chain components and design reusable LangChain workflows.

In the next chapter, we'll apply these concepts to a practical use case: building intelligent chatbots with LangChain.

CHAPTER 4

Building Chatbots

In this chapter, we will learn how to build chatbots using LangChain. We will begin by understanding why LangChain is well-suited for conversational AI and how to model conversation flows. Through hands-on examples, we will create simple chatbots, add context awareness, and handle complex queries and multi-turn interactions.

Why Use LangChain for Chatbots?

LangChain is a robust framework for developing AI applications that require conversational ability. Unlike traditional chatbot frameworks, LangChain integrates deeply with large language models (LLMs) and enables features such as memory, context awareness, and the seamless integration of external tools like APIs and databases. These features make LangChain an ideal choice for building intelligent, dynamic, and adaptable chatbots.

Key Advantages

1. Context Management

LangChain's memory modules allow chatbots to maintain conversation context over multiple turns, enhancing the user experience. For instance, in a customer service chatbot, context awareness ensures the bot remembers details like the user's name or prior queries. This creates a seamless and personalized interaction.

Code Example:

```
from langchain.chains import ConversationChain
from langchain.memory import ConversationBufferMemory
from langchain.chat_models import ChatOpenAI

llm = ChatOpenAI(temperature=0.7, model="gpt-3.5-turbo")
```

CHAPTER 4 BUILDING CHATBOTS

```
memory = ConversationBufferMemory()
chatbot = ConversationChain(llm=llm, memory=memory)

# Simulate a multi-turn conversation
print(chatbot.run("Hi, my name is Alice."))
print(chatbot.run("Can you help me with my account?"))
```

Output:

Bot: Hello Alice! How can I assist you with your account?

Here, the chatbot remembers the user's name and maintains context across turns.

This code creates a conversational chatbot using LangChain's ConversationChain, OpenAI's GPT model, and a memory buffer. It maintains context across multiple user inputs, allowing for coherent multi-turn interactions. The ConversationBufferMemory stores previous exchanges, enabling the bot to remember the user's name and respond contextually to follow-up queries.

2. Tool Integration

LangChain's ability to integrate with APIs, databases, and other external tools makes it a versatile choice for handling complex workflows. For example, a travel assistant chatbot can query APIs for flight schedules, hotel bookings, or weather updates, for example, integrating a weather API with LangChain:

```
from langchain.tools import Tool
from langchain.agents import initialize_agent, AgentType
from langchain.chat_models import ChatOpenAI

def get_weather(location):
    return f"The weather in {location} is sunny and 25°C."

weather_tool = Tool(
    name="Weather",
    func=get_weather,
    description="Provides weather updates for a given location."
)
```

```
llm = ChatOpenAI(temperature=0.7, model="gpt-3.5-turbo")
agent = initialize_agent([weather_tool], llm, agent_type=AgentType.
CONVERSATIONAL_REACT_DESCRIPTION)
print(agent.run("What's the weather in Paris?"))
```

Output:

Bot: The weather in Paris is sunny and 25°C.

This code defines a custom weather tool and integrates it into a LangChain agent using the CONVERSATIONAL_REACT_DESCRIPTION agent type. The agent leverages an LLM to interpret the user's query, decides to use the get_weather tool, and generates a natural language response. It enables dynamic tool use within a conversational interface.

3. Modular Architecture

LangChain's modular design allows developers to customize components like chains, memory, and tools. This flexibility makes it easier to build tailored solutions for specific use cases.

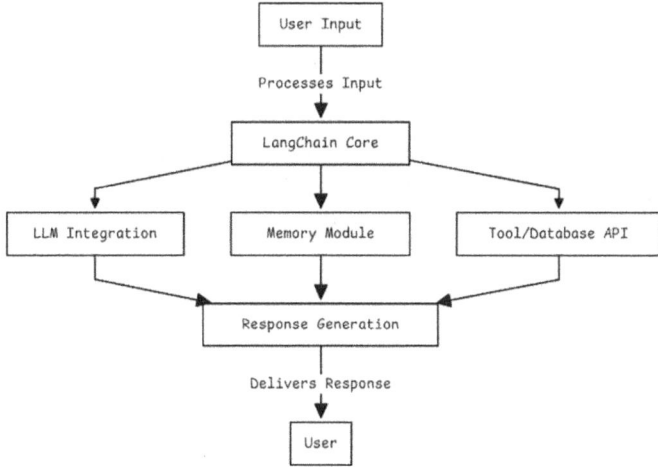

Figure 4-1. Process flow from user input to response generation for chatbot

4. Scalability

LangChain's components, such as agents and chains, are highly scalable. Developers can start with a simple chatbot and incrementally add features such as complex tool integrations, advanced memory, and fine-tuned LLMs without significant rework. For example, Scaling from Simple to Complex:

- **Initial Bot:** Handles basic FAQs using predefined templates
- **Advanced Bot:** Incorporates APIs, dynamic memory, and multi-turn dialogues

5. Support for Multiple LLMs

LangChain supports a variety of LLMs, allowing developers to switch models or combine multiple models based on performance and cost considerations. For example, a chatbot could use GPT-4 for complex queries and a lightweight model for simpler tasks.

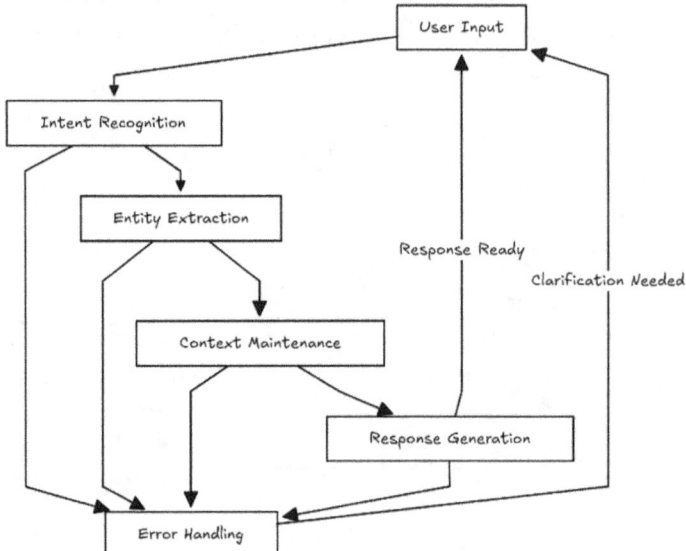

Figure 4-2. Workflow for user input handling process with contextual response generation

Understanding Conversation Flows

Within frameworks like LangChain, managing conversation flows is streamlined through its modular architecture, which allows developers to plug in specialized components like memory, chains, and language models. This modularity determines how the bot interacts with users, processes inputs, and delivers responses. Designing effective conversation flows ensures that chatbots provide meaningful, coherent, and user-friendly interactions.

Components of a Conversation Flow

1. **User Input:** The starting point of any conversation. This can range from simple text queries to voice commands or other forms of input like button clicks or forms.

2. **Intent Recognition:** The chatbot identifies the purpose of the user's input, such as asking for information, executing a task, or resolving an issue. Intent recognition is often powered by Natural Language Processing (NLP) models.

3. **Entity Extraction:** Extracting specific pieces of information from the user's input. For example, in a query like "What's the weather in Paris?", the entity is "Paris."

4. **Context Maintenance:** Storing relevant information from the conversation to maintain continuity. For instance, if a user says, "Book a flight," followed by "to New York," the bot should remember "flight" as the context for "to New York."

5. **Response Generation:** Crafting a response based on the user's input, extracted entities, and context. This response can be dynamic (generated using an LLM) or static (retrieved from predefined templates).

6. **Error Handling:** Managing unexpected inputs or failures gracefully by prompting the user for clarification or offering alternative solutions.

Conversation Flow Design

When designing a conversation flow, it's essential to

1. **Define the Use Case:** Determine the chatbot's purpose, such as customer support, lead generation, or personal assistance.

2. **Identify Key Intents:** List the primary actions users will perform, such as checking account balances, booking appointments, or troubleshooting issues.

3. **Map Out Possible Interactions:** Use flow diagrams to visualize all potential user-bot interactions.

4. **Incorporate Feedback Loops:** Ensure the bot can ask clarifying questions or confirm information when needed.

Best Practices for Conversation Flows

1. **Keep It Natural:** Use conversational language that aligns with the user's expectations.

2. **Ensure Context Awareness:** Utilize memory to provide continuity and relevance.

3. **Anticipate User Needs:** Predict follow-up questions or actions to streamline the experience.

4. **Handle Errors Gracefully:** Provide fallback responses and ask clarifying questions when needed.

5. **Iterate Based on Feedback:** Continuously refine the flow using user feedback and testing.

Building a Simple Chatbot with LangChain

Building a chatbot from scratch may seem daunting, but LangChain provides tools and abstractions to streamline the process. This section demonstrates how to create a basic chatbot that can process user queries and generate meaningful responses.

Step-by-Step Guide to Build a Simple Chatbot

1. Define the Chatbot's Purpose

Start by identifying the chatbot's main objective. For this example, we'll create a general-purpose chatbot capable of answering user queries conversationally.

2. Set Up LangChain Environment

Ensure you have the necessary dependencies installed:

```
pip install langchain openai
```

3. Initialize the Language Model

LangChain integrates seamlessly with OpenAI's GPT models. You can configure the model based on your requirements:

```python
from langchain.chat_models import ChatOpenAI

# Initialize the model
def create_model():
    llm = ChatOpenAI(temperature=0.7, model="gpt-3.5-turbo")
    return llm

llm = create_model()
```

This code defines a function to initialize and return a ChatGPT model (gpt-3.5-turbo) with a specified creativity level (temperature=0.7). It encapsulates model creation for reuse and modularity, allowing consistent setup of the language model across different parts of an application.

4. Add Memory for Context Maintenance

Memory allows the chatbot to remember past interactions, essential for multi-turn conversations:

```python
from langchain.memory import ConversationBufferMemory

# Configure memory
memory = ConversationBufferMemory()
```

5. Create the Chatbot Chain

The ConversationChain module ties the language model and memory together to handle user interactions:

```python
from langchain.chains import ConversationChain

# Create conversation chain
chatbot = ConversationChain(llm=llm, memory=memory)
```

6. Build the Chat Interface

Design a simple interface to interact with the chatbot:

```python
# Chat interface
def simple_chat():
    print("Hello! I am your chatbot. Type 'exit' to quit.")
    while True:
        user_input = input("You: ")
        if user_input.lower() == "exit":
            print("Goodbye!")
            break
        response = chatbot.run(user_input)
        print(f"Bot: {response}")

simple_chat()
```

Output:
Hello! I am your chatbot. Type 'exit' to quit.
You: What is LangChain?
Bot: LangChain is a framework for building applications with large language models. It helps manage components like memory, tools, and chains.
You: Can it remember our conversation?
Bot: Yes, I can maintain context using memory. For example, I know you just asked about LangChain.
You: That's impressive!
Bot: Thank you! Let me know if you have more questions.

This code implements a simple command-line chat interface using a LangChain-powered chatbot. It continuously accepts user input, uses the chatbot.run() method to generate context-aware responses, and prints them. The conversation persists across turns thanks to integrated memory, enabling the chatbot to remember and refer to earlier messages.

Chatbot Architecture

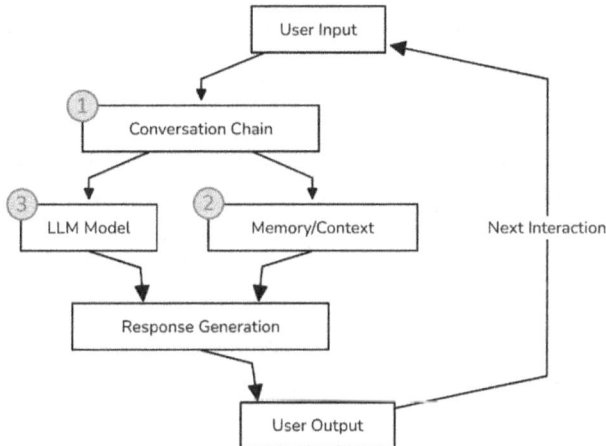

Figure 4-3. *Conversation flow in LangChain with conversation chain, memory, LLM, and response generation*

Explanation

1. **Conversation Chain:** Serves as the orchestrator, connecting the user input, memory, and the language model

2. **Memory:** Stores past interactions to provide context for the current conversation

3. **LLM:** Processes the input and generates relevant responses

7. Customizing the Chatbot

To tailor the chatbot for specific tasks:

1. **Add Custom Prompts:** Modify the initial prompt to align with the chatbot's use case:

CHAPTER 4 BUILDING CHATBOTS

```
chatbot = ConversationChain(
    llm=llm, memory=memory,
    verbose=True
)
chatbot.prompt.template = "You are a helpful assistant specialized in answering tech-related queries."
```

This modified ConversationChain sets a custom prompt to guide the chatbot's behavior, instructing it to act as a tech-savvy assistant. By updating the prompt.template, the bot tailors its responses to tech-related queries while still maintaining conversation history using `ConversationBufferMemory`. The verbose=True flag enables logging for easier debugging and traceability.

2. **Integrate External APIs:** Enhance functionality by integrating APIs. For example, retrieve weather data:

```
from langchain.tools import Tool
from langchain.agents import initialize_agent, AgentType
from langchain.chat_models import ChatOpenAI

def get_weather(location):
    return f"The weather in {location} is sunny and 25°C."

weather_tool = Tool(
    name="Weather",
    func=get_weather,
    description="Provides weather updates for a given location."
)

# llm = ChatOpenAI(temperature=0.7, model="gpt-3.5-turbo")
agent = initialize_agent([weather_tool], llm, agent_type=AgentType.CONVERSATIONAL_REACT_DESCRIPTION)

print(agent.invoke("What's the weather in Paris?"))
```

This foundational approach can be extended to create more advanced and specialized bots. The above code creates a simple AI agent using LangChain that can answer weather-related queries. A custom tool get_weather is wrapped using LangChain's Tool class and integrated into a conversational agent. The agent uses an LLM to interpret natural language, decide when to call the tool, and respond appropriately based on the tool's output.

Implementing Context Awareness in Conversations

Context awareness is a critical feature for chatbots, enabling them to maintain information across multiple turns of conversation and deliver more meaningful and personalized responses. LangChain provides robust tools and memory modules to implement context awareness seamlessly.

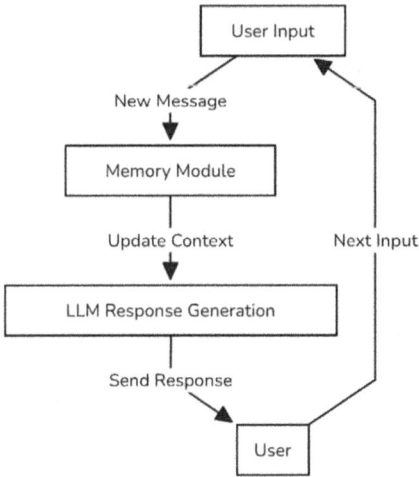

Figure 4-4. *User interaction flow with memory*

Why Context Awareness Matters

Without context, chatbots would treat each user's input as an isolated query, resulting in repetitive or irrelevant responses. For example:

- **Without Context Awareness**
  - **User:** "Book a flight to New York."
  - **Bot:** "Flight booked."

- User: "What about my hotel?"
- Bot: "I don't understand."
- **With Context Awareness**
  - User: "Book a flight to New York."
  - Bot: "Flight booked. Do you also need a hotel?
  - User: "Yes."
  - Bot: "Hotel booked in New York."

Best Practices for Context-Aware Chatbots

- **Limit Memory Scope:** Avoid overloading memory with unnecessary details to optimize performance.
- **Combine Memory Types:** Use a mix of buffer and summary memory to handle complex interactions.
- **Test for Edge Cases:** Ensure the bot handles incomplete or ambiguous inputs gracefully.

Handling Complex Queries and Multi-turn Dialogues

Building chatbots capable of handling complex queries and managing multi-turn dialogues is essential for creating advanced conversational systems. LangChain provides the tools and modularity required to efficiently implement these features.

Complex Queries

Complex queries often involve multiple sub-tasks, require additional data retrieval, or include ambiguous intent. To handle such scenarios effectively, a chatbot must

1. **Break Down Queries:** Decompose user queries into manageable sub-tasks.
2. **Use External Tools:** Fetch data or interact with APIs to gather required information.

CHAPTER 4 BUILDING CHATBOTS

3. **Generate Coherent Responses:** Synthesize a comprehensive reply that integrates all results.

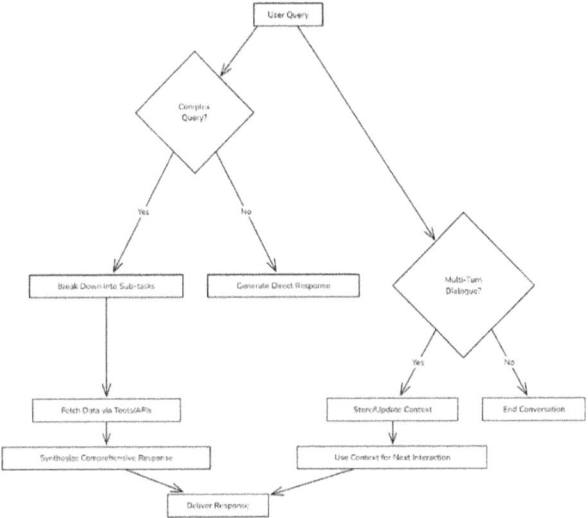

Figure 4-5. Decisionflow for handling simple vs. complex user queries

Example: Travel Assistant Query

User Input: "Find me flights to Paris and a hotel near the Eiffel Tower."

1. **Break Down Intent:** Identify two sub-tasks: (a) search for flights to Paris, and (b) search for nearby hotels.

2. **Data Retrieval:** Use APIs for flights and hotels.

3. **Response Synthesis:** Combine results into a single, coherent response.

Code Implementation

```
from langchain.tools import Tool
from langchain.agents import initialize_agent, AgentType
from langchain.chat_models import ChatOpenAI

# Define mock tools for flights and hotels
def find_flights(destination):
    return f"Flights to {destination}: AirFrance, Delta, and Lufthansa."
```

127

```
def find_hotels(location):
    return f"Hotels near {location}: Hotel A, Hotel B, and Hotel C."
flight_tool = Tool(
    name="Flight Finder",
    func=find_flights,
    description="Finds flights to a specified destination."
)
hotel_tool = Tool(
    name="Hotel Finder",
    func=find_hotels,
    description="Finds hotels near a specified location."
)
# Initialize LangChain agent
llm = ChatOpenAI(temperature=0.7, model="gpt-3.5-turbo")
agent = initialize_agent([flight_tool, hotel_tool], llm, agent_type=AgentType.CONVERSATIONAL_REACT_DESCRIPTION)
# Handle the query
query = "Find me flights to Paris and a hotel near the Eiffel Tower."
response = agent.run(query)
print(response)
```

Output:

Flights to Paris: AirFrance, Delta, and Lufthansa.
Hotels near the Eiffel Tower: Hotel A, Hotel B, and Hotel C.

This code sets up a LangChain agent equipped with two custom tools for finding flights and hotels. When given a travel-related query, the agent uses an LLM to interpret the request, intelligently invoke the relevant tools, and return a coherent, multi-part response. It demonstrates tool orchestration within a natural language interface.

Multi-turn Dialogues

Multi-turn dialogues involve maintaining context across multiple interactions. LangChain's memory modules enable chatbots to

1. **Store Context:** Remember user inputs, intents, and entities.
2. **Update Context Dynamically:** Modify context based on new user inputs.
3. **Provide Relevant Follow-ups:** Use stored context to deliver meaningful responses.

Example: Banking Assistant

Sample user interaction:

User: "What's my balance?"
Bot: "Your balance is $1,500."
User: "Transfer $500 to John."
Bot: "Transfer to John confirmed. Your new balance is $1,000."

Code Implementation

```
from langchain.chains import ConversationChain
from langchain.memory import ConversationBufferMemory
from langchain.chat_models import ChatOpenAI

# Initialize LLM and memory
llm = ChatOpenAI(temperature=0.7, model="gpt-3.5-turbo")
memory = ConversationBufferMemory()

# Create a conversation chain
banking_bot = ConversationChain(llm=llm, memory=memory)

# Simulate a multi-turn dialogue
print(banking_bot.run("What's my balance?"))
print(banking_bot.run("Transfer $500 to John."))
```

Output:

```
Bot: Your balance is $1,500.
Bot: Transfer to John confirmed. Your new balance is $1,000.
```

This code demonstrates a banking chatbot using LangChain's ConversationChain with memory. The ConversationBufferMemory enables the bot to retain context across multiple user inputs, allowing it to simulate tasks like checking balances and transferring money while maintaining state between turns for a realistic, context-aware conversation.

Best Practices

1. **Define Sub-task Handling Logic:** Plan how to decompose complex queries effectively.

2. **Ensure Context Accuracy:** Verify that stored context aligns with user inputs.

3. **Graceful Error Handling:** Manage incomplete or ambiguous queries by prompting users for clarification.

4. **Optimize External Calls:** Minimize latency when fetching data from APIs or databases.

By integrating the ability to manage complex queries and multi-turn dialogues, you can build chatbots that are not only more capable but also offer a highly engaging and seamless user experience.

Key Takeaways

- LangChain simplifies chatbot development by combining chains, memory, and tools.
- You can model conversation flows and multi-turn logic.
- Memory and context management are key to coherent interactions.
- Advanced use cases include complex queries and decision trees.
- LangChain agents can be embedded into chatbots for more autonomy.
- You'll gain hands-on experience building your own conversational AI application.

With a solid chatbot foundation, the next chapter dives into the Retrieval-Augmented Generation (RAG) pipeline to build knowledge-grounded applications.

CHAPTER 5

Building Retrieval-Augmented Generation (RAG) Systems

This chapter provides a comprehensive guide to designing and building Retrieval-Augmented Generation systems using LangChain. We will explore each stage of the RAG pipeline from loading and preprocessing data, chunking and embedding, indexing into vector stores, to retrieval and final response generation. Practical techniques for improving retrieval quality and ensuring ethical, explainable behavior are also covered.

Overview of the RAG

Retrieval-Augmented Generation (RAG) represents a paradigm shift in how large language models (LLMs) interact with information. At its core, RAG combines the power of retrieval systems with the generative capabilities of LLMs to produce responses that are both contextually relevant and factually grounded.

Unlike traditional LLMs which rely solely on their parametric knowledge—information encoded during training—RAG enables models to access and leverage external knowledge sources at inference time. This hybrid approach allows the model to "look up" information before generating a response, much like a human might consult reference materials when faced with a specific question.

The benefits of RAG are multifaceted and significant:

- **Enhanced Accuracy and Trustworthiness:** By grounding responses in retrieved documents, RAG significantly reduces hallucinations—those plausible-sounding but factually incorrect outputs that plague standard LLMs.

- **Knowledge Recency:** RAG systems can access up-to-date information beyond a model's training cutoff, keeping responses current without requiring retraining.

- **Domain Adaptation:** Organizations can inject domain-specific knowledge into general-purpose LLMs, customizing them for specialized applications without expensive fine-tuning.

- **Transparency and Auditability:** RAG provides clear provenance for information, allowing users to trace assertions back to source documents—a critical feature for applications requiring explainability.

- **Reduced Deployment Costs:** For many applications, RAG offers a more cost-effective alternative to fine-tuning or training domain-specific models from scratch.

As AI systems become increasingly integrated into knowledge-intensive workflows across industries, RAG has emerged as a foundational technique that bridges the gap between the computational efficiency of LLMs and the factual reliability demanded by real-world applications.

Note In this chapter, rather than dissecting individual code blocks in every section, we will present a comprehensive case study. This approach will illuminate the underlying concepts of RAG and its various stages, providing a holistic understanding of the system.

Components of a RAG System

Building a RAG system comprises multiple interconnected components:

- **Retriever:** Uses dense, sparse, or hybrid methods to fetch context.
- **Generator:** Synthesizes responses via LLMs (e.g., GPT-4, Claude).
- **Knowledge Source:** Vector stores (Pinecone), document DBs, or APIs.
- **Pipeline Orchestrator:** LangChain chains manage retrieval ➤ generation workflow.

Now let's go deeper into each component.

Retriever

The retriever is responsible for gathering relevant information from external sources. Key techniques involved include

- **Dense Retrieval**: Utilizes embeddings and vector similarity search to find relevant data

- **Sparse Retrieval**: Relies on traditional keyword matching methods, such as BM25, to retrieve information

- **Hybrid Retrieval**: Combines both dense and sparse retrieval methods to improve the accuracy and efficiency of information retrieval

- **Neural Retrieval**: Leverages deep learning models, such as transformers, to capture semantic relationships between query and document

- **Contextualized Retrieval**: Incorporates context, such as user history or session data, to enhance the relevance of retrieved results

- **Query Expansion**: Enhances the original query by adding related terms or synonyms to improve the retrieval quality

- **Knowledge Graph-Based Retrieval**: Uses a graph of interconnected data points to enhance the retrieval process, focusing on relationships between entities

Generator

The generator is responsible for synthesizing coherent and contextually relevant responses based on the retrieved information. Key examples and techniques include

- **LLMs (Large Language Models)**: Powerful models such as OpenAI's GPT, Anthropic's Claude, and Meta's LLaMA, which generate contextually appropriate and coherent responses

- **Fine-Tuned Models**: Models that have been specifically adapted or trained on domain-specific data or for particular use cases to improve their performance in each context

- **Neural Text Generation**: Leveraging deep learning architectures to produce fluent and meaningful text based on the input data

- **Conditional Text Generation**: Generates text conditioned on specific input or context, ensuring responses align with provided information or requirements

- **Transformer-Based Models**: Employs the transformer architecture to model long-range relationships and produce contextually nuanced text

- **Zero-Shot or Few-Shot Models**: Models capable of generating responses for untrained tasks by leveraging preexisting knowledge to deduce suitable answers

- **Reinforcement Learning from Human Feedback (RLHF)**: Enhances response generation by integrating human feedback to improve quality, relevance, and safety

- **Encoder-Decoder Models**: Uses a dual architecture where the encoder analyzes input data and the decoder produces corresponding output based on that analysis

Knowledge Source

A knowledge source refers to the storage and organization of information that can be retrieved during processing. Key types of knowledge sources include

- **Vector Store**: Specialized databases, such as Pinecone or Weaviate, that store and retrieve data based on vector representations for efficient similarity search and retrieval

- **Document Store**: Includes unstructured text files, such as PDFs or Word documents, as well as structured sources like relational databases that organize and store text and data for easy access and retrieval

- **Graph Database**: A type of database that uses graph structures to store and manage data, emphasizing relationships between entities, which can be particularly useful for knowledge graphs

- **Knowledge Graph**: A network of interconnected entities and relationships, offering a semantic layer that enhances data retrieval based on context and connections between concepts

- **File System**: Traditional storage systems that manage file-based data, such as cloud storage or local disk systems, which can be indexed for retrieval

- **Cloud Data Stores**: Distributed cloud platforms like AWS, Azure, or Google Cloud that manage large-scale data storage, often integrating with other retrieval techniques for dynamic access

- **Content Management Systems (CMS)**: Platforms used to manage, store, and organize large volumes of content or documents, facilitating easy access and retrieval for specific applications

Pipeline Orchestrator

LangChain's pipeline orchestrator ties the retriever and generator into a seamless workflow.

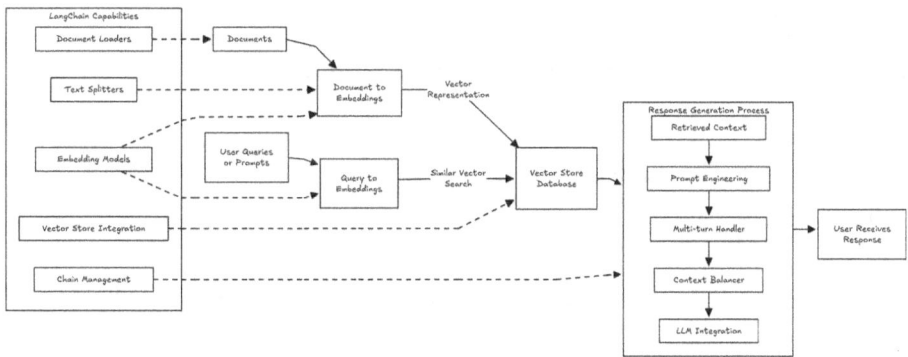

Figure 5-1. *Architecture of a Retrieval-Augmented Generation (RAG) system*

The diagram illustrates the comprehensive architecture of a RAG (Retrieval-Augmented Generation) system implemented using LangChain capabilities. It shows the end-to-end flow from document ingestion to user response generation; refer to Figure 5-1.

LangChain's core components:

1. **Document Loaders**: Tools that import documents from various sources
2. **Text Splitters**: Components that segment documents into manageable chunks
3. **Embedding Models**: Services that convert text into vector representations
4. **Vector Store Integration**: Connectors to various vector databases
5. **Chain Management**: Orchestration layer that coordinates the flow between components

Document Processing Pipeline

The diagram shows how documents flow through the system:

- Document Loaders feed into Documents.
- Documents, along with Text Splitters and Embedding Models, feed into the "Document to Embeddings" process.
- This process converts documents into vector representations.

Query Processing Pipeline

Similarly for user queries:

- User Queries/Prompts are processed into vector embeddings using the same Embedding Models.
- The "Query to Embeddings" block transforms user input into the same vector space as the documents.

Vector Storage and Retrieval

- Both document and query embeddings feed into the Vector Store/Database.
- Similar Vector Search connects queries to relevant document vectors.

Response Generation Process

The diagram details the sophisticated response generation pipeline:

1. **Retrieved Context**: Relevant document chunks from the vector search
2. **Prompt Engineering**: Formulation of effective prompts that combine the query with retrieved context
3. **Multi-turn Handler**: Management of conversation history and context
4. **Context Balancer**: Optimization of context window usage
5. **LLM Integration**: Connection to the language model that generates the final response

The diagram also shows that Chain Management can directly influence the Response Generation Process, allowing for customized workflows.

The final output is delivered to the user as a coherent response that integrates retrieved information with the language model's capabilities.

Approach to RAG Implementation

Implementing Retrieval-Augmented Generation (RAG) with LangChain involves several key steps to set up a robust system for retrieving and generating contextually relevant responses. Here's a breakdown of the process.

Step 1: Define the Knowledge Source
The first step in implementing RAG is to define the knowledge source that will be used for information retrieval. This source can be a variety of data types, depending on the use case. Common options include indexed documents stored in vector databases or external APIs that provide real-time data. The knowledge source serves as the foundation for retrieving relevant information when a query is processed.

Step 2: Select the Retriever
Once the knowledge source is defined, the next step is to select the retriever. The retriever is responsible for fetching the relevant information based on the input query. LangChain provides modules to facilitate this, such as the FAISS (Facebook AI Similarity Search) vector store, which can store vectors representing documents and perform efficient similarity searches. The retriever can be configured by loading embeddings (such as OpenAI embeddings) and setting up the retriever to access the vector store.

Step 3: Configure the Generator

With the retriever set up, the next step is to configure the generator. The generator, typically a large language model (LLM), is responsible for generating a coherent and contextually accurate response based on the information retrieved by the retriever. LangChain integrates with LLMs, such as OpenAI's GPT models, allowing easy setup for the text generation process.

Step 4: Orchestrate the Pipeline

The retriever and generator are then combined into a cohesive pipeline. This pipeline ensures smooth interaction between the two components. LangChain's **RetrievalQA** chain can be used to seamlessly orchestrate the process, where a query is first processed by the retriever to fetch relevant information, and then the generator uses that information to create a response. Once the pipeline is configured, you can run it by passing a query and receiving the generated response.

Step 5: Evaluate and Optimize

After setting up the pipeline, it's important to evaluate the performance of the RAG system. Metrics such as accuracy, relevance, and latency are used to assess how well the system retrieves and generates responses. LangChain's LangSmith can be utilized for debugging, monitoring, and fine-tuning the system to improve its overall performance. This ensures that the system remains accurate, responsive, and efficient over time.

By following these steps, you can implement a powerful RAG system using LangChain, capable of retrieving relevant information and generating high-quality responses tailored to user queries.

Use Cases and Applications of RAG

Retrieval-Augmented Generation (RAG) has diverse applications across various industries, where it enhances the performance of systems by combining information retrieval with generative capabilities. Some key use cases include

- **Customer Support Automation**

    RAG can be used in customer service chatbots and virtual assistants. The retriever fetches relevant knowledge from customer support documents, product manuals, and FAQs, while the generator creates personalized, context-aware responses, improving customer experience and reducing response times.

- **Legal Document Review**

   In the legal industry, RAG can assist lawyers in reviewing contracts, case laws, and legal documents. By retrieving relevant sections or precedents and generating concise summaries or insights, RAG helps lawyers save time while ensuring they have the most relevant information at their fingertips.

- **Healthcare and Medical Research**

   RAG is valuable in healthcare for summarizing medical research, retrieving relevant clinical guidelines, or providing insights from patient records. It can assist healthcare professionals by generating evidence-based responses, diagnoses, or treatment recommendations while referencing the latest research and case studies.

- **Personalized Recommendations**

   RAG can enhance recommendation systems in e-commerce, entertainment, or education platforms. By retrieving past user interactions, reviews, and preferences, it can generate tailored suggestions for products, movies, or courses, leading to a more personalized and engaging user experience.

- **Content Generation and Curation**

   Content creators, journalists, and marketers can leverage RAG for generating articles, blog posts, or product descriptions. The retriever fetches relevant data from articles, papers, and databases, while the generator creates human-like content based on the retrieved information, enhancing productivity and creativity.

- **Finance and Market Analysis**

   In the financial sector, RAG can assist analysts by retrieving relevant financial reports, market data, and news articles and then generating insights, trend analyses, or forecasting reports. This allows for quicker decision-making and better-informed strategies in investment or risk management.

- **Education and Tutoring**

    RAG can be applied to create intelligent tutoring systems. The retriever can fetch educational materials such as textbooks, research papers, or answers from online resources, while the generator helps create personalized explanations, quizzes, and interactive learning sessions.

- **News Aggregation and Analysis**

    RAG is useful in news aggregation tools, where it retrieves the latest news articles or reports from multiple sources and generates concise summaries. It can also generate analysis, helping users stay up to date on trends and developments across various fields.

Data Loading and Preprocessing

LangChain supports a wide range of data sources and formats, enabling flexible integration into various workflows. Key supported sources include

File-Based Data Sources

- Text files (.txt)
- PDFs (.pdf)
- CSVs (.csv)

Database and Storage

- SQL databases (MySQL, PostgreSQL)
- NoSQL databases (MongoDB)
- Cloud storage systems (AWS S3, Azure Blob Storage)

Web and APIs

- Web scraping for dynamic content
- RESTful APIs for real-time data

LangChain provides adapters for seamless ingestion of these data formats.

CHAPTER 5 BUILDING RETRIEVAL-AUGMENTED GENERATION (RAG) SYSTEMS

Techniques for Efficient Data Loading

Efficient data loading is crucial for ensuring smooth performance in systems that process large datasets. Two common strategies for optimizing data loading are

1. **Batch Loading**

    Loading data in smaller, manageable chunks helps prevent memory overflow and enhances performance. LangChain supports batch loading by enabling document splitting, making it easier to process large datasets in chunks without overwhelming memory.

2. **Parallel Processing**

    Parallelization allows multiple data loading tasks to be processed simultaneously, significantly speeding up the ingestion of large datasets. LangChain integrates with Python's parallel processing tools, allowing efficient parallel loading of documents across multiple threads or processes.

Both strategies improve performance and ensure that the system can handle large volumes of data effectively.

Data Cleaning and Normalization

Before feeding data into a Retrieval-Augmented Generation (RAG) system, it's essential to clean and normalize the data to ensure consistency and relevance. This process involves several key steps:

1. **Text Cleaning**

    Removing unnecessary elements like special characters, HTML tags, or stopwords helps improve the quality of the data. Additionally, normalizing the text by converting it to lowercase ensures uniformity across the dataset. This process enhances the accuracy of the retrieval and generation phases of RAG.

    LangChain helps by providing built-in tools for text splitting and processing, which can easily integrate with custom cleaning functions.

2. **Duplicate Removal**

   Redundant or duplicate entries can lead to inaccurate or biased results. Removing duplicates ensures that only unique and relevant documents are considered during the retrieval process, improving retrieval accuracy.
   LangChain does not directly handle duplicate removal, but it can be integrated with existing Python methods like set() to remove duplicates from the dataset.

3. **Tokenization**

   Tokenization involves breaking down the text into smaller, manageable pieces (tokens) that are compatible with embeddings. This is crucial for transforming text into a format that can be processed by models for efficient retrieval and generation.
   LangChain provides an easy way to tokenize and split documents using tools like CharacterTextSplitter, which can segment documents into tokens for further processing.

By cleaning, normalizing, and tokenizing the data, we ensure that the RAG system operates with high-quality, relevant, and consistent input, leading to more accurate and effective outputs.

Handling Different Types of Data (Text, PDFs, Web Content)

LangChain offers specialized modules for handling different types of data, making it easier to load and process content from various sources. Here's how LangChain helps with different data types:

1. **Text Files**

   LangChain provides a simple and direct way to load text files, making it easy to work with plain text data. The TextLoader module is used to read and load text files into a format that can be processed further.

   LangChain's TextLoader module allows straightforward loading of text files, providing easy integration for document processing.

CHAPTER 5 BUILDING RETRIEVAL-AUGMENTED GENERATION (RAG) SYSTEMS

2. **PDFs**

    For PDF documents, LangChain includes the PyPDFLoader module, which efficiently parses and extracts text content from PDFs. This is especially useful when dealing with structured data that may be stored in PDF format.

    LangChain's PyPDFLoader efficiently extracts content from PDFs, making it accessible for downstream processing in the RAG system.

3. **Web Content**

    LangChain can dynamically scrape content from web pages using the WebBaseLoader module. This allows for the extraction of relevant data from live websites, providing real-time information for the system.

    LangChain's WebBaseLoader enables web scraping and content extraction directly from web pages, integrating real-time web data into the RAG pipeline.

With these specialized modules, LangChain simplifies the process of loading, parsing, and processing content from different data sources, making it easier to integrate diverse forms of information into a RAG system.

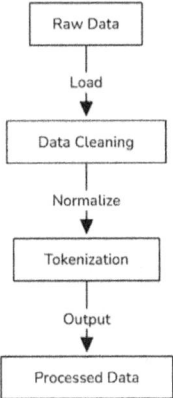

Figure 5-2. Data loading workflow

Here we have explored how to preprocess the data (refer to Figure 5-2), i.e., to ingest, preprocess, and normalize data effectively for use in RAG pipelines. These steps are foundational to ensuring high-quality retrieval and generation outcomes.

Chunking Strategies

Text chunking is a critical preprocessing step in building RAG systems. It involves splitting large documents or data sources into smaller, manageable pieces (chunks) to enable efficient retrieval and relevance scoring. Effective chunking ensures that the retriever retrieves only the most relevant and concise information, which in turn enhances the generator's output.

Why Chunking Matters

- **Improves Retrieval Accuracy**: Smaller chunks make it easier to match queries with relevant portions of the text.

- **Reduces Computational Overhead**: By working with smaller data units, the system can operate faster and more efficiently.

- **Enhances Contextual Relevance**: Proper chunking ensures that retrieved segments are self-contained and coherent.

Fixed-Length Chunking

Fixed-length chunking is a technique used to divide a document into segments of equal length, where each chunk contains a predefined number of characters or tokens. This method is straightforward and easy to implement, making it suitable for tasks where the chunks are required to be uniform in size. However, it does not consider the meaning or structure of the text, which can sometimes lead to fragmented or semantically broken pieces of information.

The main idea behind fixed-length chunking is to split a document into smaller pieces (chunks), each of which contains a specific number of characters or tokens. This is useful in cases where uniformity is more important than understanding the meaning of the content within each chunk. For instance, when processing large amounts of text or preparing data for machine learning, chunks of a fixed size can simplify the process.

Benefits

1. **Simplicity**: Fixed-length chunking is simple to implement and doesn't require sophisticated logic.

2. **Uniformity**: It ensures that each chunk is the same size, which can be useful when a specific input length is required by downstream processes (e.g., feeding data into machine learning models that require fixed-size inputs).

Limitations

1. **Semantic Loss**: Since the chunking process does not take into account the structure of the text, it can split important context, like a sentence or paragraph, into separate chunks. This can lead to chunks that are harder to understand or process correctly.

2. **No Context Preservation**: It doesn't maintain semantic relationships between words across chunks, which could be crucial in certain NLP tasks such as summarization or translation.

Semantic Chunking Techniques

Semantic chunking aims to divide text based on its meaning rather than just its length or structure. The goal is to ensure that each chunk represents a logically coherent and contextually complete segment of information. This approach is ideal when you want the chunks to preserve some semantic boundaries, such as grouping text around topic shifts, headers, or paragraph structures.

Concept

- **Purpose**: The focus is on splitting text in a way that maintains the semantic integrity of each segment. For example, you might split a document at the end of paragraphs, sections, or after a specific topic shift to ensure that chunks are meaningful and coherent.

- **How It Works**: Instead of just using a fixed length, semantic chunking uses natural text markers (e.g., paragraph breaks, headers, or punctuation) to guide the chunking process.

Benefits

1. **Preserves Meaning**: Each chunk will contain contextually related content, which makes it easier to understand and process.

2. **Suitable for Complex Documents**: This approach works well for documents that are divided into distinct topics, sections, or paragraphs, where the logical flow matters.

Sentence- and Paragraph-Based Chunking

This technique splits the text at natural linguistic boundaries, such as sentences or paragraphs, to preserve the semantic coherence of each chunk. It ensures that chunks are not arbitrarily cut, which helps maintain the overall structure and meaning of the text.

Concept

- **Purpose:** The text is divided based on sentences or paragraphs, which naturally form meaningful units. This approach helps ensure that chunks remain understandable and contextually coherent.

- **How It Works**: Sentences or entire paragraphs are treated as the smallest units of meaningful content. When the text is split, it keeps the integrity of each sentence or paragraph intact.

Benefits

- **Retains Semantic Meaning:** Since chunks are split at natural language boundaries (sentences or paragraphs), the meaning is preserved.

- **Ideal for Narrative or Structured Text:** This method is especially useful for documents like stories, reports, or articles, where preserving the structure of sentences and paragraphs is crucial.

Overlapping Chunks and Sliding Windows

Overlapping chunking involves creating chunks that share some content between adjacent segments. This technique helps maintain context across chunks and is especially useful in tasks like information retrieval, where knowing the relationship between chunks can improve accuracy.

Concept

- **Purpose**: By ensuring that consecutive chunks overlap, this method allows important context from one chunk to be carried over into the next. This reduces the likelihood of losing context when text is divided into smaller parts.

- **How It Works**: The sliding window technique creates chunks that overlap by a specified amount (e.g., 100 characters), ensuring that the end of one chunk is repeated at the beginning of the next chunk.

Benefits

- **Context Preservation:** Overlapping chunks help maintain context between consecutive segments, which is crucial for tasks like question answering or text retrieval.

- **Improves Retrieval Accuracy:** Since adjacent chunks share content, there is a better chance that relevant information will be found across multiple chunks, enhancing the effectiveness of queries.

Optimizing Chunk Size for Different Use Cases

When working with large text datasets in tasks such as retrieval-augmented generation (RAG) or information retrieval, **chunk size** plays a pivotal role in balancing **context preservation** and **retrieval efficiency**. The optimal chunk size will depend on the specific requirements of your task, such as the level of detail needed in responses, the scope of queries, and the amount of context a chunk should hold.

Here's how different chunk sizes impact performance and usability.

Smaller Chunks

- **Suitable for Granular Queries**: Smaller chunks (e.g., 200 characters) are useful when queries require very specific, precise matches. For example, when searching for a particular fact or a short piece of information, smaller chunks make it easier to locate the exact content.

- **Benefit**: They enable faster retrieval and more accurate pinpointing of the relevant information.

- **Limitation**: However, smaller chunks often lose important context, making it harder to comprehend the full meaning without additional chunks being retrieved. This can negatively impact tasks that require a broader understanding.

Larger Chunks

- **Better for Broad Queries**: Larger chunks (e.g., 1000 characters or more) are more appropriate when the query needs broader context or when the content in the document is complex and interconnected.

- **Benefit**: They preserve more context and are suitable for tasks like summarization or question answering, where understanding the full context of the document is important.

- **Limitation**: The trade-off with larger chunks is slower retrieval time and potentially larger memory usage, as each chunk is more substantial.

Key Considerations for Chunk Size Selection

1. **Context vs. Efficiency**
   - Smaller chunks improve query speed but may lose context.
   - Larger chunks preserve more context but may result in slower retrieval.

2. **Task Type**
   - Granular tasks (e.g., fact retrieval) benefit from smaller chunks.
   - Tasks requiring comprehension of larger documents (e.g., summarization) benefit from larger chunks.

3. **Overlap Impact**: The amount of overlap between chunks also affects context retention. Overlap helps ensure that each chunk doesn't lose important information, especially for larger chunks.

4. **Document Length**: Longer documents may require larger chunk sizes to preserve meaningful context, while shorter documents can often be divided into smaller chunks without losing important details.

Embedding Data

Introduction to Text Embeddings

Text embeddings are numerical representations of text data, capturing semantic meaning in a vector format. These dense, high-dimensional vectors enable machines to process and compare textual information efficiently. Embeddings are foundational to RAG systems as they enable similarity-based retrieval, clustering, and semantic search.

Importance of Embeddings in RAG

- **Semantic Understanding**: Embeddings represent textual content in a way that semantically similar texts are closer in vector space.

- **Efficient Retrieval**: Vectorized data allows fast similarity searches using techniques like k-Nearest Neighbors (kNN).

- **Model-Agnostic Integration**: Generative models can use embeddings to focus on relevant contextual data.

Embedding Models Supported in LangChain

LangChain integrates with several popular **embedding models**, both open source and proprietary, allowing users to choose the best model for their specific task. Embedding models are essential for transforming text data into vector representations that capture semantic meaning, enabling efficient retrieval, search, and various NLP applications. Here are some of the embedding models supported in LangChain.

1. OpenAI Embeddings

- **Example:** text-embedding-ada-002
- **Description:** OpenAI's embeddings, such as text-embedding-ada-002, are highly flexible and suitable for a wide variety of tasks. These embeddings can capture nuanced semantic information, making them ideal for general-purpose applications.
- **Best For**
  - **General-Purpose Embeddings:** Suitable for tasks like text classification, similarity search, and clustering
  - **Versatility:** Effective across different domains without the need for domain-specific tuning
- **Use Cases:** Question answering, semantic search, document retrieval, summarization.

2. Sentence Transformers

- **Example:** all-MiniLM-L6-v2
- **Description:** Sentence Transformers are known for generating high-quality sentence embeddings that capture semantic meaning. The all-MiniLM-L6-v2 model is one of the most popular variants and works well for tasks that require understanding the relationship between different sentences or texts.
- **Best For**
  - **Semantic Search:** Optimized for tasks that involve searching for semantically similar sentences or text segments.
  - **Clustering Tasks:** Useful in grouping similar content based on meaning.
- **Use Cases:** Information retrieval, clustering, text similarity, semantic search engines.

3. Cohere Embeddings

- **Example**: embed-english-light-v2.0

- **Description**: Cohere provides high-performance embeddings that are particularly effective with multilingual text. They offer models like embed-english-light-v2.0, which provide fast and accurate embeddings, making them suitable for a wide variety of applications.

- **Best For**

    - **Multilingual Text**: Especially suitable for tasks where text is in multiple languages

    - **High Performance**: Known for producing embeddings quickly and efficiently

- **Use Cases**: Cross-lingual search for documents, multilingual NLP applications, general text embedding tasks.

4. Hugging Face Models

- **Example: sentence-transformers library** (for custom and domain-specific embeddings)

- **Description**: Hugging Face offers a rich ecosystem of models, including the **sentence-transformers** library, which provides pre-trained models for generating sentence and text embeddings. These models are highly customizable, allowing for domain-specific fine-tuning to produce embeddings that are more relevant for specialized use cases.

- **Best For**

    - **Custom Embeddings**: Ideal for creating domain-specific embeddings tailored to your particular application.

    - **Custom Fine-Tuning**: You can fine-tune models on your own dataset for more accurate results.

- **Use Cases**: Domain-specific applications, text classification, similarity comparison, custom NLP solutions.

Generating Embeddings for Chunks

To effectively retrieve relevant data from large datasets, each chunk of text must be converted into an **embedding vector**. These vectors represent the semantic meaning of the text, enabling efficient search and retrieval. LangChain simplifies this process by providing easy-to-use tools and integration with various embedding models.

Example Workflow

1. **Split Data into Chunks**: Use a chunking strategy, as discussed in the previous section, to break down the data into smaller, manageable pieces (chunks).

2. **Generate Embeddings for Each Chunk**: Once the data is split into chunks, apply the selected embedding model to convert each chunk into an embedding vector.

Generating Embeddings for Chunks

- **Chunking**: Text data is first divided into smaller pieces, or "chunks," which can be processed individually.

- **Embedding Generation**: After chunking, an embedding model is applied to each chunk to convert it into a vector representation. These embeddings capture the semantic meaning of the text.

- **Vector Store Indexing**: The generated embeddings are then indexed using a vector store (e.g., FAISS). This enables fast retrieval of relevant chunks when performing searches or other tasks that rely on semantic similarity.

Handling Large-Scale Embedding Tasks

For large datasets, generating embeddings can be computationally intensive. LangChain provides tools and techniques to optimize this process and make it more efficient.

Batch Processing

Batch processing allows the dataset to be processed in smaller batches, which helps avoid memory bottlenecks and reduces the load on computational resources. Instead of processing all chunks at once, the data is split into manageable subsets, and each batch is processed sequentially.

Benefit: Batch processing makes it possible to handle large datasets by processing smaller portions at a time, thereby reducing memory consumption and computational overhead.

Distributed Computing

For very large datasets, distributed computing frameworks like **Dask** or **Apache Spark** can be used to parallelize the embedding generation process. These frameworks allow the workload to be distributed across multiple machines or cores, drastically improving processing speed and scalability.

Benefit: Distributed computing enables faster embedding generation by utilizing the full power of multiple machines or processors, making it suitable for extremely large datasets.

Persistent Storage for Embeddings

Once embeddings are generated, they can be stored in persistent storage systems such as **Pinecone**, **Weaviate**, **Qdrant**, or **Milvus**. These databases are optimized for storing and retrieving embeddings efficiently, allowing for scalable and fast retrieval without having to regenerate embeddings each time a query is made.

Benefit: Persistent storage solutions provide an efficient way to store embeddings for quick access during retrieval tasks, which is particularly useful for large-scale systems.

Key Techniques for Large-Scale Embedding Generation

1. **Batch Processing**
   - Reduces memory load by processing smaller chunks of data at a time
   - Ideal for scenarios where data is too large to process in a single pass

2. **Distributed Computing**

    - Speeds up embedding generation by parallelizing the work across multiple machines or cores
    - Suitable for handling vast datasets that would be too time-consuming for a single machine

3. **Persistent Storage**

    - Embeddings are stored in optimized databases like Pinecone, Weaviate, or Milvus for fast retrieval.
    - Ideal for applications that require frequent access to embeddings without needing to regenerate them.

Indexing: Vector Stores

Vector databases form the backbone of RAG systems by enabling efficient storage, retrieval, and management of vectorized data. Unlike traditional databases, vector databases are designed to work with high-dimensional embeddings, allowing similarity-based queries such as k-Nearest Neighbors (kNN).

Key Benefits of Vector Databases

1. **Scalability**: Handle millions of embeddings efficiently.
2. **Fast Retrieval**: Perform similarity searches in real-time.
3. **Integration with ML Workflows**: Seamlessly work with embeddings generated by models.

Types of Vector Stores Supported in LangChain

LangChain supports a variety of vector stores, catering to different requirements such as open source solutions and cloud-based services. These include

1. **FAISS**: Ideal for small-to-medium datasets with local deployment
2. **Pinecone**: Best for production systems needing managed services

3. **Weaviate**: Suitable for applications requiring hybrid search

4. **Chroma**: Lightweight, developer-friendly option for local experimentation

Creating and Managing Vector Indices

Vector indices allow for efficient querying of embedding data by organizing vector representations in a way that enables fast similarity searches. In Retrieval-Augmented Generation (RAG) systems, these indices help retrieve relevant information from a knowledge base. LangChain provides built-in support for various vector databases, simplifying index creation and management.

For example, FAISS (Facebook AI Similarity Search) is a commonly used library for local experimentation. It stores high-dimensional vectors and enables fast approximate nearest neighbor searches. LangChain integrates seamlessly with FAISS and other vector databases, making it easy to create, store, and retrieve embeddings efficiently.

Scalability and Performance Considerations

When dealing with large-scale applications, scalability and performance optimization become crucial. Key strategies include

1. **Sharding and Partitioning**: Splitting data across multiple nodes to distribute the workload

2. **GPU Acceleration**: Utilizing GPUs for faster vector similarity computations

3. **Index Compression**: Reducing memory footprint using techniques like Product Quantization (PQ)

For large datasets, cloud-based vector stores like Pinecone or Milvus provide scalability, ensuring fast and efficient retrieval in distributed environments. LangChain supports these platforms, allowing developers to build scalable RAG pipelines.

Choosing the Right Vector Store for Your Application

Selecting an appropriate vector store depends on various factors:

- **High Scalability and Reliability**: Pinecone, Milvus
- **Local Experimentation**: FAISS, Chroma

- **Hybrid Search Capabilities**: Weaviate, Qdrant
- **Cost Efficiency**: FAISS, Chroma

Decision Framework

1. **Dataset Size**: Small datasets work well with FAISS, while large ones benefit from distributed stores like Pinecone or Milvus.
2. **Query Latency**: Applications requiring real-time responses might need optimized solutions like Pinecone or Milvus.
3. **Ease of Use**: Weaviate and Pinecone offer user-friendly APIs for quick integration.

LangChain abstracts vector database operations, providing an easy-to-use interface for indexing and querying embeddings. By leveraging LangChain's integrations, developers can seamlessly implement vector search in RAG pipelines without deep knowledge of underlying vector database implementations.

Retrieval Techniques

Similarity Search Algorithms

Similarity search algorithms are essential in retrieval systems within RAG (Retrieval-Augmented Generation). These algorithms compare query embeddings with stored embeddings to find the most relevant information. LangChain supports multiple similarity search methods, each suited for different use cases:

- **Euclidean Distance:** Measures the straight-line distance between vectors. It works well in low-dimensional spaces but is less effective in high-dimensional embeddings.
- **Cosine Similarity:** Measures the angle between two vectors, making it ideal for text embeddings where direction matters more than magnitude.

- **Approximate Nearest Neighbor (ANN):** Speeds up retrieval by approximating nearest neighbors instead of performing exact searches. Algorithms like Hierarchical Navigable Small World (HNSW) are commonly used for efficiency.

LangChain integrates these similarity search methods, allowing seamless implementation in RAG pipelines.

Dense Retrieval vs. Sparse Retrieval

Retrieval methods are classified as dense or sparse (see Table 5-1 for a summary), and LangChain supports both types, allowing for flexible search options.

- **Dense Retrieval:** Dense retrieval methods encode data as high-dimensional embeddings within a continuous vector space. This approach excels at capturing the semantic relationships between queries and documents, enabling more accurate and context-aware search results. However, dense retrieval typically demands significant computational resources, as it relies on specialized vector search infrastructure to efficiently compare and retrieve relevant items from large datasets.

- **Sparse Retrieval:** Sparse retrieval methods utilize traditional keyword-based techniques such as TF-IDF (Term Frequency-Inverse Document Frequency) and BM25. These methods represent documents as sparse vectors, where only a small subset of terms are non-zero. Sparse retrieval is highly efficient in terms of computation and storage, making it well-suited for large-scale search applications. However, its reliance on exact keyword matching limits its ability to understand the deeper semantic meaning of queries and documents, often resulting in less relevant results when synonyms or related concepts are involved.

Table 5-1. Comparison of Dense vs. Sparse Retrieval

Feature	Dense Retrieval	Sparse Retrieval
Semantic Search	Yes	Limited
Speed	Slower (without ANN)	Faster
Data Type	Unstructured	Structured
Tools	Embedding Models	TF-IDF, BM25

Hybrid Retrieval Approaches

Hybrid retrieval combines dense and sparse methods to leverage their strengths. A typical workflow involves

1. Using sparse retrieval to filter out irrelevant documents

2. Applying dense retrieval to rank and refine results based on semantic similarity

This approach is particularly useful for applications that require both keyword precision and semantic relevance, such as legal document search or customer support chatbots. LangChain provides built-in support for hybrid retrieval, simplifying implementation.

Implementing Custom Retrieval Methods

LangChain allows developers to create custom retrieval methods tailored to specific needs. This may involve

- Integrating external APIs (e.g., enterprise search engines)
- Implementing domain-specific ranking algorithms
- Applying additional filtering logic based on business rules

A custom retriever in LangChain can be implemented by defining a class that inherits from its base retriever, ensuring flexibility in retrieval strategies.

Metadata Filtering and Faceted Search

To improve retrieval accuracy, metadata filtering and faceted search help refine search results:

- **Metadata Filtering:** Adds constraints based on metadata fields like author, category, or date
- **Faceted Search:** Groups result into structured categories for better navigation, commonly used in e-commerce and document management

LangChain supports these filtering techniques, enabling precise and user-friendly retrieval experiences.

Improving Model Retrieval

Improving model retrieval focuses on refining how a system retrieves and ranks documents in a Retrieval-Augmented Generation (RAG) pipeline. Enhanced retrieval leads to more accurate and contextually relevant outputs. Below is a detailed explanation of each component along with illustrative examples.

Query Expansion and Reformulation

This process involves modifying the user's original query to capture the full scope of its intent. The goal is to broaden or fine-tune the query so that the system retrieves a more comprehensive and relevant set of documents.

Synonym Expansion
By adding synonymous terms, the system can cover variations in language.
Example: A user searches for "fast car." By expanding the query, the system also includes terms such as "quick automobile" and "speedy vehicle." This helps capture documents that might use different words to describe the same concept.

Reweighting Keywords
Certain keywords might be more important in a given domain. Adjusting the weights means that these terms are emphasized more during the retrieval process.

Example: In a medical context, if a query includes the word "treatment," that term might be assigned more importance than a generic word like "health," ensuring that the retrieved documents focus on actionable medical information.

Rephrasing

Reformulating a query can help clarify ambiguity or align with the language used in the target documents.

Example: A query such as "Apple stock" could be ambiguous—it might refer to the technology company or the fruit. Rephrasing it to "stock market performance of Apple Inc." helps the system focus on financial information.

Re-ranking Retrieved Documents

Re-ranking involves adjusting the order of documents after the initial retrieval phase to ensure that the most relevant and high-quality documents appear at the top of the results.

Score Aggregation

Multiple retrieval methods (such as dense and sparse retrieval) may produce different relevance scores for the same document. Aggregating these scores (for instance, by averaging or weighted summing) can produce a more balanced final ranking.

Example: If one method indicates a document has a high relevance score but another method assigns a lower score, combining the scores can moderate the ranking, leading to a final order that reflects overall relevance more accurately.

Model-Based Re-ranking

Beyond similarity scores, additional signals like document recency, author reputation, or user click behavior can be used to re-rank documents.

Example: Imagine a scenario where two documents are semantically similar to the query. One comes from a reputable source, while the other is from a less credible site. A model-based re-ranking approach would prioritize the credible source.

Metadata-Based Re-ranking

Leveraging metadata (such as publication dates, document types, or user ratings) can further refine the ranking process.

Example: In an academic research system, documents published in peer-reviewed journals might be ranked higher than those from non-reviewed sources, even if their initial relevance scores are similar.

Relevance Feedback Mechanisms

Relevance feedback enables the system to learn and improve over time by incorporating user feedback on retrieved documents. This process can be based on either explicit feedback (direct user input) or implicit feedback (inferred from user behavior).

Explicit Feedback

Users directly rate the usefulness of a document.

Example: After reading a retrieved article, a user might be prompted to rate its helpfulness on a scale of 1 to 5. The system can use these ratings to adjust future rankings for similar queries.

Implicit Feedback

The system observes user behavior, such as which documents are clicked on or how long a user spends on a page, to infer relevance.

Example: If users consistently click on a particular document or spend significant time reading it, the system learns that the document is likely relevant and can boost its ranking in subsequent queries.

Fine-Tuning Retrieval Models

Fine-tuning involves adapting pre-trained models to perform better within a specific domain or context. This process is essential for handling specialized vocabulary and nuanced concepts. Here are the steps:

1. **Prepare Training Data**

    Collect domain-specific queries paired with relevant documents.
    Example: In a legal application, a dataset could include queries related to legal cases and corresponding case summaries.

2. **Select a Pre-trained Model**

    Begin with a general-purpose model that understands language well, such as models from SentenceTransformers or OpenAI.
    Example: A pre-trained model that has been effective in general natural language understanding can be chosen as the starting point.

3. **Train the Model**

   Adapts the transformer architecture through training on domain-specific datasets to better capture the intricacies, specialized terminology, and contextual nuances of a particular field. For example, fine-tuning on legal documents enables the model to master legal jargon, understand critical case details, and enhance its accuracy and relevance in legal research scenarios.

4. **Integrate into LangChain**

   Once fine-tuned, the specialized model replaces the generic retrieval model in the system, leading to more accurate and context-aware retrieval.

   Example: In a customer support system, a model fine-tuned on past support tickets and resolutions can more effectively match user queries to the correct troubleshooting articles.

Ensemble Methods for Improved Retrieval

Ensemble methods combine the strengths of multiple retrieval approaches, resulting in a more robust system that can handle diverse queries and data types.

Techniques

Weighted Voting

Different retrieval models contribute to the final result based on their assigned weights, reflecting their reliability or performance in a given context.

Example: A system might use a dense retrieval model weighted at 60% and a sparse retrieval model weighted at 40%. The final ranking of documents is determined by combining the scores from both models according to these weights.

Stacked Models

Outputs from multiple retrieval systems are used as inputs for a secondary model that learns to rank the documents optimally.

Example: The secondary model may analyze the relevance scores from several retrieval systems along with additional metadata (like document length or recency) to produce a final, improved ranking.

Diverse Retrieval
Combining dense, sparse, and hybrid retrieval approaches can capture both semantic meaning and exact keyword matches, ensuring comprehensive search results.

Example: For a complex query, the dense retrieval method might capture subtle semantic relationships, while the sparse method ensures that specific keywords are matched. The ensemble then integrates both aspects for a balanced output.

LangChain serves as a modular framework that integrates these advanced retrieval techniques seamlessly. It provides pre-built components for query preprocessing, retrieval, re-ranking, feedback incorporation, and model fine-tuning. This modularity allows developers to mix and match different approaches according to their application's needs.

Example in Context

Consider a customer support system where users report technical issues. LangChain can preprocess the query (expanding and rephrasing terms like "laptop overheating" to include related phrases such as "thermal issues" and "device cooling"), retrieve a broad set of relevant documents using both dense and sparse methods, and then re-rank these results based on user feedback and metadata like document recency. Over time, the system can be fine-tuned using support ticket data, and ensemble methods can combine various retrieval strategies to ensure the most helpful information is always presented at the top.

Response Generation Using LLMs

Response generation is the final step in a RAG system, where a language model produces an answer by integrating user queries with retrieved context. The quality of the final response depends on how well the retrieved information is combined with the model's inherent knowledge. Below is a detailed explanation of the key components and techniques.

CHAPTER 5   BUILDING RETRIEVAL-AUGMENTED GENERATION (RAG) SYSTEMS

Integrating Retrieved Context with LLM Prompts

Integrating retrieved context involves taking information from a retrieval system—such as relevant documents or text chunks—and embedding it within a prompt for the language model. This ensures that the LLM is provided with the necessary background to generate accurate and context-aware responses.

How It Works

1. **Retrieval:** Start by querying a vector database or search system to gather relevant documents or text passages based on the user's query.

2. **Preprocessing:** Extract or summarize the key information from the retrieved content.

3. **Prompt Structuring:** Combine the retrieved context with the user query in a well-defined template that instructs the LLM on how to use the context.

Imagine a user asks, "What is the capital of France?" The system retrieves a document containing "The capital of France is Paris." The prompt then explicitly presents the context along with the query, so the LLM understands where to focus, such as

- **Context:** "The capital of France is Paris."
- **Question:** "What is the capital of France?"
- **Instruction:** "Use the above context to provide an answer."

This structured approach helps the LLM produce a precise and informed response.

Prompt Engineering for RAG Systems

Concept Overview

Prompt engineering involves designing the prompt in a way that guides the LLM to utilize the retrieved context effectively while minimizing hallucinations and ensuring the answer remains grounded.

Strategies and Examples

- **Instructional Prompts**

  Clearly instruct the model on how to incorporate the provided context.

  Example: "You are an AI assistant. Use only the context below to answer the question. If the answer is not in the context, say 'I don't know.'"

- **Few-Shot Examples**

  Including a few examples of similar queries and ideal responses can help the LLM understand the desired format and depth.

  Example

  Context: "The capital of Italy is Rome."

  Question: "What is the capital of Italy?"

  Answer: "Rome."

- **Constraints**

  Set explicit boundaries for the LLM, so it prioritizes the retrieved context and avoids incorporating unrelated information.

  Example: "Only use the given context to answer the following query."

These strategies ensure that the prompt is clear, focused, and effective in guiding the model.

Handling Multi-turn Conversations in RAG

Multi-turn conversations require maintaining context over several interactions. The system must integrate both new queries and past interactions to ensure continuity and coherence in the responses.

Approaches

- **History Summarization**

    Instead of including every previous message, summarize the key points of earlier interactions to reduce token usage while preserving context.

    Example: "Previously, we discussed that the Eiffel Tower is in Paris and was designed by Gustave Eiffel."

- **Dynamic Context Updating**

    Append only the most relevant parts of the conversation history to the prompt alongside the new query.

    Example: If the user's query builds on an earlier topic, include a brief recap: "Earlier, you mentioned details about Paris landmarks. Now, you ask about the construction date of the Eiffel Tower."

- **Context Window Management**

    If the conversation is long, older messages might be truncated or summarized to stay within token limits without losing essential context.

By carefully managing the conversation history, the system maintains a coherent thread across multiple exchanges.

Balancing Retrieved Information and Model Knowledge

A critical challenge in RAG is balancing the explicit retrieved context with the model's pre-trained knowledge. Relying too heavily on one source might either lead to incomplete answers or unwanted hallucinations.

Strategies

- **Weighting Retrieved Context**

    Structure the prompt to emphasize the retrieved context.

    Example: "Use the following context to answer the question. If the context is insufficient, you may use your own knowledge, but indicate the uncertainty."

- **Relevance Filters**

    Ensure that only highly relevant information is included in the context so that the model's own general knowledge only supplements, rather than overrides, the retrieved data.

- **Fallback Mechanisms**

    In cases where the retrieved context is sparse or ambiguous, the prompt may instruct the model to rely on its internal knowledge.

    Example: "If the retrieved information does not cover the answer, provide your best guess based on general knowledge."

Techniques for Maintaining Coherence and Relevance

Maintaining coherence and relevance in the final response is essential for usability and accuracy. This involves careful prompt design and iterative refinement of outputs.

Techniques

- **Chunk Prioritization**

    Order the retrieved chunks so that the most relevant or authoritative information is placed first in the prompt.

    Example: For a query about historical events, ensure that the prompt starts with a chronologically accurate and highly relevant passage.

- **Post-Processing**

    After the LLM generates an answer, additional processing can be applied to refine and polish the response. This may include grammatical corrections, reordering sentences for clarity, or verifying key facts.

- **Feedback Loops**

    Incorporate user feedback to continuously improve the prompt and retrieval process.

    Example: If users frequently indicate that certain responses are off-topic, the system can adjust the prompt template to better emphasize context.

Example in Practice

In a customer support scenario, a user might ask, "How do I reset my router?" LangChain retrieves troubleshooting documents and user guides, integrates this information into a structured prompt, and uses prompt engineering techniques to ensure the LLM generates a clear, accurate response that references the specific troubleshooting steps, while also handling follow-up queries in a coherent conversation.

Ethical Considerations and Best Practices

Ethical considerations are essential in designing and deploying RAG systems. The following sections outline key areas where ethical integrity must be maintained, along with best practices and examples to illustrate each concept.

Handling Sensitive Information in RAG Systems

Sensitive information can be inadvertently retrieved or generated by RAG systems, posing risks such as privacy violations or misuse of data. It is critical to implement strategies that prevent sensitive data from being exposed.

- **Content Filtering**

  Before indexing data, apply preprocessing techniques to identify and redact sensitive information. For example, systems can scan for terms like "SSN," "credit card," or "password" and replace them with placeholder text. This ensures that any sensitive details do not get stored or used during retrieval.

- **Access Control**

  Implement strict controls to limit access to sensitive datasets. Only authorized users or systems should be able to view or modify this data, minimizing the risk of accidental exposure.

- **Ethical Guidelines**

  Develop and enforce clear ethical guidelines that dictate how sensitive information should be handled both during retrieval and in the generated outputs. These guidelines serve as a framework for developers and users to ensure responsible usage.

Example

In a financial application, if a query inadvertently includes sensitive data such as credit card details, a robust system would automatically detect and redact this information before it is processed or indexed.

Bias Mitigation in Retrieval and Generation

Bias in RAG systems can arise from the training data, indexing process, or even the language model itself. Unchecked bias may result in unfair or unbalanced responses.

Techniques for Bias Mitigation

- **Balanced Training Data**

  Ensure that the data used for both training and indexing is diverse and represents various perspectives. This helps reduce the likelihood of skewed responses.

- **Bias Detection Tools**

  Use specialized tools to analyze and detect biases in the outputs of the retrieval and generation processes. Once detected, these biases can be corrected through adjustments in the model or the training process.

- **Contextual Adjustments**

  Fine-tune prompts and retrieval methods to avoid stereotypical or biased associations. For example, when retrieving information about inventors, instruct the system to include data about innovators from diverse backgrounds.

Example

If a system is asked to retrieve information about inventors, the prompt might explicitly encourage the inclusion of inventors from various cultures and regions, ensuring that the results are balanced and inclusive.

Transparency and Explainability in RAG

Transparency and explainability are crucial for building trust in RAG systems. Users should be able to understand how and why the system arrived at a particular response.

Methods for Explainability

- **Citation of Sources:** Display the documents or data chunks that were used to generate the response. This helps users verify the origin of the information and assess its credibility.

- **Explainable AI Techniques:** Use models and methods that allow insights into the decision-making process. This might include providing confidence scores or detailed breakdowns of how different pieces of information contributed to the final answer.

- **User-Friendly Interfaces:** Develop interfaces that clearly show provenance information, such as the sources used and their corresponding relevance scores, thereby offering a window into the system's internal reasoning.

Example

In an academic research tool, when a user queries a historical fact, the system might display the key source texts along with a confidence rating, indicating which sources were most influential in generating the response.

Data Privacy and Compliance Considerations

Adhering to data privacy regulations, such as GDPR or CCPA, is vital for any system handling personal data. Compliance protects user rights and helps prevent legal and reputational risks.

Key Considerations

- **Data Anonymization**

  Remove or mask personal identifiers in datasets before indexing. This step minimizes the risk of exposing personal data.

- **Audit Trails**

  Maintain detailed logs of data access and usage within the system. Audit trails ensure that every interaction is recorded, allowing for accountability and troubleshooting.

- **Consent Mechanisms**

    Implement processes that require explicit user consent for data usage. Users should have the option to view, edit, or delete their personal data, ensuring their control over sensitive information.

Example

In a healthcare information system, before data is used for training or retrieval, all personal identifiers such as emails, phone numbers, and patient IDs are anonymized. Furthermore, patients are informed about how their data will be used and must provide consent, reinforcing privacy and compliance.

Conclusion

This chapter covers the building blocks of Retrieval-Augmented Generation systems using LangChain, guiding readers from foundational concepts to advanced implementation and optimization. By blending retrieval and generation, RAG systems deliver accurate, up-to-date, and context-aware responses, addressing key limitations of standalone LLMs. The chapter equips readers with the tools and knowledge to design robust, scalable, and ethical RAG pipelines for diverse real-world applications.

Key Takeaways

- RAG systems combine retrieval of relevant documents with LLM-based response generation.
- LangChain supports loading diverse data types and chunking strategies.
- Embedding models and vector stores power similarity search.
- Advanced retrieval and ranking techniques improve performance.
- RAG pipelines require careful prompting, context integration, and tuning.
- Ethical considerations such as bias, privacy, and explainability are critical for responsible deployment.

CHAPTER 5   BUILDING RETRIEVAL-AUGMENTED GENERATION (RAG) SYSTEMS

Next, you'll explore how to deploy, optimize, and monitor language model workflows using LangServe, LangSmith, and LangGraph. These tools enable production-ready deployments, advanced debugging, and the design of complex, modular AI workflows—taking your LangChain applications from prototype to scalable, maintainable systems.

CHAPTER 6

LangServe, LangSmith, and LangGraph: Deploying, Optimizing, and Designing Language Model Workflows

As we've explored the capabilities of LangChain, vector databases, tools, and language models, it's essential to discuss how to deploy, optimize, and design workflows that unlock their full potential. This chapter delves into three critical components of the LangChain ecosystem: LangServe, LangSmith, and LangGraph.

We begin with LangServe, which enables the deployment of language models for scalable and efficient inference. By serving models through LangServe, developers deploy models in production environments.

Next, we'll explore LangSmith, a framework designed for debugging, optimizing, and fine-tuning language models. With LangSmith, we can refine their models to achieve better performance, accuracy, and reliability.

LangGraph, which facilitates the design of complex language model workflows. By leveraging LangGraph, developers can create customized workflows that integrate multiple models, data sources, and tasks. Figure 6-1 explains the ecosystem in action.

CHAPTER 6   LANGSERVE, LANGSMITH, AND LANGGRAPH: DEPLOYING, OPTIMIZING, AND DESIGNING LANGUAGE MODEL WORKFLOWS

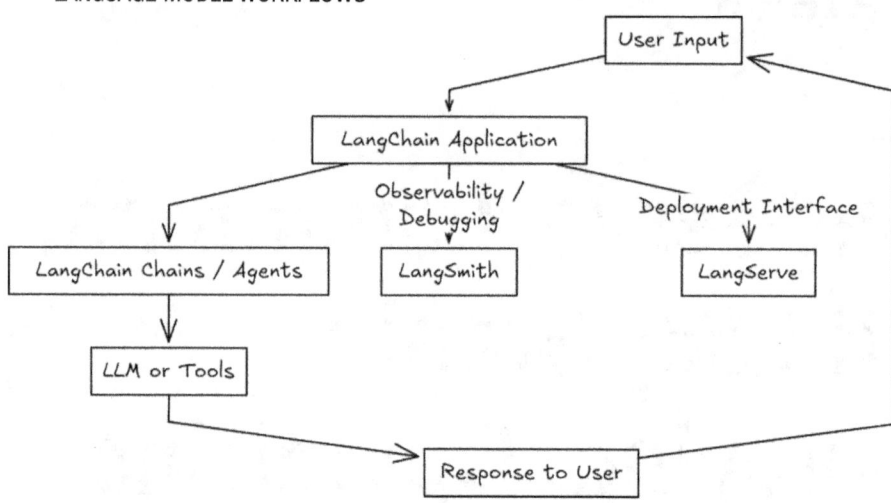

Figure 6-1. *LangChain ecosystem in action*

Feature	LangServe	LangSmith	LangGraph
Purpose	Deploy LangChain-based applications as APIs	Debugging, monitoring, and evaluation of LangChain applications	Building complex, multi-step workflows in LangChain
Key Functionality	Converts chains into FastAPI-based services, simplifies deployment, provides API endpoints for integration with other services	Provides detailed execution traces, allows for logging and visualizing intermediate steps, helps in model evaluation and debugging	Enables DAG (Directed Acyclic Graph)-based chaining, allows complex branching and parallel processing, supports structured flow control in AI applications
When to Use	When you need to expose your GenAI application as an API for external integration, when deploying LangChain models in production	When troubleshooting and optimizing the performance of LangChain applications, when you need insights into token usage, latency, and error tracking	When your GenAI workflow involves multiple steps, conditional branching, and dependencies between different models or data sources

(*continued*)

Feature	LangServe	LangSmith	LangGraph
Best For	Serving LangChain applications as APIs, seamless deployment to cloud environments	Developers and ML engineers who need deep visibility into AI workflows, debugging, logging, and performance tuning	Applications requiring structured AI workflows with dependencies and conditional logic, experimenting with different processing paths
Example Use Case	Deploying a chatbot or document summarization service as an API	Analyzing and debugging an RAG (Retrieval-Augmented Generation) pipeline, tracking chain execution and failure points	Designing a customer support chatbot that first retrieves knowledge base data, summarizes it, and then validates the response using another AI model

LangServe

LangServe is an essential tool for deploying LangChain applications, providing a streamlined approach to serving chains as APIs. It is built on FastAPI and Uvicorn, enabling developers to create scalable and production-ready inference endpoints for their LLM applications.

Uses of LangServe

LangServe plays a critical role in simplifying the deployment of LangChain applications. Here are some of its key use cases.

Custom Deployments

One of the primary uses of LangServe is deploying custom LangChain applications as APIs. By using LangServe, developers can wrap their chains into endpoints that can be accessed by external applications.

Example: Deploying a Custom Chain with LangServe

```
from langchain.chains import LLMChain
from langchain.prompts import PromptTemplate
from langchain.llms import OpenAI
from langserve import add_routes
from fastapi import FastAPI

# Define a simple chain
prompt = PromptTemplate(input_variables=["question"], template="Answer the following question: {question}")
llm = OpenAI()
chain = LLMChain(llm=llm, prompt=prompt)

# Create FastAPI app
app = FastAPI()
add_routes(app, chain, path="/generate")

# Run the server using uvicorn
# uvicorn filename:app --host 0.0.0.0 --port 8000
```

In this example, we create an API endpoint /generate that takes input and returns a response from the language model. This makes it easier to integrate the chain with web applications, chatbots, or other services.

Integration with LangChain and Other Frameworks

LangServe seamlessly integrates with LangChain, allowing developers to serve their chains effortlessly. Additionally, it can work alongside other frameworks such as Flask, Django, and cloud services like AWS Lambda.

This approach allows developers to serve multiple chains via different API routes, making it easier to manage different functionalities.

LangServe API Configuration with Observability

LangServe supports streaming responses, making it suitable for applications that require real-time text generation, such as chatbots.

```
import langfuse
```

```python
from langchain_core.runnables.config import RunnableConfig
from langfuse import Langfuse
from langfuse.langchain import CallbackHandler
from fastapi import FastAPI
from langserve import add_routes

# Initialize the main Langfuse client for auth checking
langfuse_client = Langfuse()

# Initialize the callback handler for LangChain integration
langfuse_handler = CallbackHandler()

# Test the connection using the main client
try:
    langfuse_client.auth_check()
    print("Langfuse connection successful!")
except Exception as e:
    print(f" Langfuse connection failed: {e}")

config = RunnableConfig(callbacks=[langfuse_handler])

llm_with_langfuse = llm.with_config(config)

# Setup server
app = FastAPI()

# Add Langserve route
add_routes(
    app,
    llm_with_langfuse,
    path="/test-simple-llm-call",
)
```

This code sets up the server that exposes the language model as an API endpoint. This initializes a callback handler for tracking and observability, applies this handler to a language model configuration, and then creates a REST API endpoint at "/test-simple-llm-call". When this endpoint receives requests, they'll be processed by the language model.

Advantages of Using LangServe

LangServe offers several benefits for LangChain developers:

- **Ease of Deployment**: Deploying LLM chains as APIs is straightforward.
- **Scalability**: Supports cloud deployment and containerization.
- **Streaming Support**: Enables real-time applications.
- **Multi-chain Support**: Serve multiple chains with different functionalities.
- **Integration**: Works well with various cloud platforms and frameworks.

LangGraph

LangGraph is a powerful library designed to facilitate the creation of structured, agentic workflows in LangChain applications. By leveraging graph-based execution, LangGraph allows developers to create dynamic, adaptable, and modular AI workflows that can handle complex interactions and dependencies efficiently.

Uses of LangGraph

LangGraph extends LangChain's capabilities by introducing graph-based execution, which allows developers to define workflows using nodes and edges. This is particularly useful for managing complex AI-driven applications, ensuring modularity and scalability.

Application of LangGraph with LangChain Deployments

LangGraph can be used to enhance LangChain deployments by structuring AI workflows as graphs. Instead of sequential chains, developers can define branching paths, loops, and decision-making processes within their applications.

Example: Creating a Graph-Based AI Workflow

```
from langgraph.graph import StateGraph, START, END
from langchain_openai import OpenAI
from langchain.prompts import PromptTemplate
from typing_extensions import TypedDict

# Define state using TypedDict (recommended approach)
class QueryState(TypedDict):
    query: str

def process_query(state: QueryState) -> QueryState:
    response = llm.invoke(f"Answer this question: {state['query']}")
    return {"query": response}

# Create the graph
workflow = StateGraph(QueryState)
workflow.add_node("process", process_query)
workflow.add_edge(START, "process")
workflow.add_edge("process", END)
app = workflow.compile()

# Run the graph
initial_state = {"query": "What is LangGraph?"}
result = app.invoke(initial_state)
print(result["query"])
```

This example showcases how to structure an AI pipeline using LangGraph, making workflows more maintainable and scalable than traditional sequential chains.

Building Agentic Workflows with LangGraph

LangGraph excels at managing agentic workflows by allowing developers to design systems where multiple agents interact dynamically. This is particularly useful for applications that require decision-making, multiple interacting LLMs, and adaptive reasoning.

Additional Applications of LangGraph

LangGraph is a versatile tool that extends beyond traditional AI applications. Some other use cases include as follows.

Automated Customer Support

LangGraph can be used to design customer support workflows, where multiple agents handle different aspects of a query, such as retrieving order details, providing troubleshooting steps, and escalating issues when needed.

Decision Support Systems

By integrating LangGraph with external data sources and APIs, developers can create decision support systems that analyze data, generate insights, and provide recommendations dynamically.

Intelligent Data Pipelines

LangGraph can orchestrate data processing workflows, where different nodes handle tasks like data cleansing, transformation, and model inference, enabling scalable and maintainable AI-driven pipelines.

LangSmith

LangSmith is a powerful tool designed to enhance the development, debugging, and deployment of AI applications built with LangChain. By providing observability, evaluation, and monitoring capabilities, LangSmith streamlines the workflow for AI developers, ensuring more efficient, robust, and reliable AI-driven applications.

Application of LangSmith with LangChain Deployments

LangSmith integrates seamlessly with LangChain deployments, enabling developers to

- **Monitor and Debug AI Workflows**: LangSmith allows developers to track execution paths, understand reasoning patterns, and debug issues in LangChain applications.
- **Optimize Prompt Engineering**: Developers can experiment with different prompts, measure their effectiveness, and iterate quickly to improve model responses.

- **Evaluate Performance Metrics**: LangSmith provides evaluation tools to assess the quality of AI-generated outputs using various metrics, including accuracy and relevance.

- **Enhance Version Control and Experiment Tracking**: Developers can store and compare different versions of AI workflows to analyze improvements over time.

This example demonstrates how LangSmith helps track agent execution and monitor its interactions in real-time, improving debugging and performance assessment.

Streamlining AI Development with LangSmith

LangSmith simplifies AI development by offering tools that enhance the overall development cycle:

- **Logging and Observability**: Captures each step of a LangChain application's execution.

- **Error Handling**: Detects failure points and logs error messages.

- **Automated Testing**: Provides built-in evaluation metrics to test AI workflows before deployment.

- **Data Insights and Analytics**: Helps developers analyze user interactions and optimize system performance.

Setting Up and Managing Projects with LangSmith

LangSmith enables efficient project management by organizing AI workflows into structured projects:

- **Creating a New Project**: Developers can group related LangChain components under a single project.

- **Managing API Keys and Access Controls**: Securely manage API keys and access permissions for team collaboration.

- **Tracking Model Variants**: Compare different versions of an AI model within a project.

CHAPTER 6   LANGSERVE, LANGSMITH, AND LANGGRAPH: DEPLOYING, OPTIMIZING, AND DESIGNING LANGUAGE MODEL WORKFLOWS

LangServe, LangSmith, and LangGraph collectively empower developers to operationalize LangChain applications—from deploying models at scale to debugging their behavior and structuring sophisticated, agent-driven workflows.

Key Takeaways

- LangServe simplifies deployment of LangChain apps as scalable, production-ready APIs.

- LangSmith provides powerful debugging, observability, and experiment tracking for LangChain workflows.

- LangGraph enables structured, stateful, agent-based workflows using graph-based execution.

- Together, these tools ensure scalability, reliability, and maintainability in enterprise GenAI systems.

In the following chapter, we will explore how to prepare and format data for fine-tuning and how LangChain can assist in this process—including managing memory, compute constraints, and integrating your fine-tuned models into LangChain workflows for seamless execution.

CHAPTER 7

LangChain and NLP

Natural Language Processing (NLP) lies at the heart of most language model applications, and LangChain offers a flexible framework to integrate both traditional and modern NLP techniques into your workflows. In this chapter, we explore how to leverage LangChain for essential NLP tasks such as sentiment analysis, text classification, and task-specific fine-tuning.

NLP Techniques in LangChain

As far as LangChain is considered, Natural Language Processing (NLP) is at the heart of it's functionality that enables applications to understand, process, and generate human language. LangChain helps developers with a framework to leverage various NLP techniques through its simple yet powerful abstractions. Let us see how LangChain implements fundamental NLP techniques and how we can utilize them effectively.

LangChain helps in development by providing a standardized way to chain together different language processing components. Instead of writing complex integration code between different NLP models and services, it offers a consistent interface that allows developers to focus on solving business problems rather than technical integration challenges.

The framework's strength lies in its ability to abstract away the complexity of working with different language models while maintaining flexibility.

Sentiment Analysis and Classification

Sentiment analysis and text classification are important NLP tasks for understanding and categorizing content. This library makes implementing these capabilities straightforward using LLMs or traditional ML models.

CHAPTER 7   LANGCHAIN AND NLP

Sentiment analysis is an important analysis for many businesses to understand customer feedback, social media mentions, product reviews, and support tickets. Traditional rule-based approaches often fall short because they can't understand context, sarcasm, or nuanced expressions. LLM-based approach leverages the contextual understanding capabilities to provide more accurate sentiment detection which is now made easier with LangChain.

Basic Sentiment Analysis Implementation

Let's start with a simple sentiment analysis chain that demonstrates LangChain's core concepts:

```
from langchain_openai import ChatOpenAI
from langchain_core.prompts import ChatPromptTemplate
from langchain_core.output_parsers import StrOutputParser

# Simple sentiment analysis chain
sentiment_prompt = ChatPromptTemplate.from_template(
    "Analyze the sentiment of the following text. Respond with only
    'positive', 'negative', or 'neutral'.\n\nText: {text}"
)

llm = AzureOpenAI(deployment_name="dp-gpt-35-turbo-instruct", model_name="gpt-35-turbo-instruct")

sentiment_chain = sentiment_prompt | llm | StrOutputParser()

# Example usage
text = "This is a great product! I love it."
sentiment = sentiment_chain.invoke({"text": text})
print(f"Sentiment: {sentiment}")
```

Breaking Down the Components

1. **ChatPromptTemplate:** This creates a reusable prompt template with placeholders. The {text} placeholder allows us to dynamically insert different text for analysis while maintaining consistent instructions.

2. **LLM Integration:** The AzureOpenAI instance connects to Azure's OpenAI service. LangChain supports multiple LLM providers, making it easy to switch between services without changing your application logic.

3. **Output Parsing:** StrOutputParser() ensures the LLM's response is returned as a clean string, removing any formatting artifacts.

Building a Text Classifier

LangChain Simplifies Creating Multi-class Text Classifiers

We can go beyond simple sentiment analysis and create a multi-class classifier for customer support scenarios:

```
from langchain.chains import LLMChain
from langchain_core.prompts import PromptTemplate

# Multi-class classification template
classification_template = """
Classify the following text into one of these categories:
- Product Question
- Technical Support
- Billing Issue
- Feature Request
- Complaint

Text: {text}

Classification (respond with only the category name):
"""

classification_prompt = PromptTemplate(
    input_variables=["text"],
    template=classification_template
)

classifier_chain = LLMChain(llm=llm, prompt=classification_prompt)

llm = AzureOpenAI(deployment_name="dp-gpt-35-turbo-instruct", model_name="gpt-35-turbo-instruct")
```

```
# Example usage
category = classifier_chain.run("My subscription was charged twice this
month.")
print(f"Category: {category}")
```

Key Design Considerations

- **Clear Categories:** The prompt explicitly lists all possible categories, helping the LLM make consistent classifications.

- **Specific Instructions:** Asking for "only the category name" prevents verbose responses that would complicate downstream processing.

- **Real-World Application:** This classifier could route customer inquiries to appropriate support teams automatically.

Advanced Classification with Structured Output

For production applications, we often need more than just a category label. We want confidence scores, alternative classifications, or additional metadata:

```
from langchain.output_parsers import PydanticOutputParser
from pydantic import BaseModel, Field
from typing import List, Dict

# Use built-in float instead of typing.Float
class ClassificationResult(BaseModel):
    primary_category: str = Field(description="The main category of
    the text")
    confidence: float = Field(description="Confidence score between
    0 and 1")  # Changed typing.Float to float
    secondary_categories: List[Dict[str, float]] = Field(  # Changed
    typing.Float to float
        description="Other possible categories with confidence scores"
    )

classification_parser = PydanticOutputParser(pydantic_
object=ClassificationResult)

advanced_classification_prompt = PromptTemplate(
    template="Classify the following text:\n{text}\n{format_instructions}",
```

```
    input_variables=["text"],
    partial_variables={"format_instructions":
    classification_parser.get_format_instructions()}
)

advanced_classifier = advanced_classification_prompt | llm |
classification_parser

# Example usage:
user_input = "I'm having trouble logging into my account. I think my
password might be incorrect."

# Invoke the advanced classifier
result = advanced_classifier.invoke({"text": user_input})

# Print the classification results
print(result)

# Access specific fields of the result
print(f"Primary Category: {result.primary_category}")
print(f"Confidence: {result.confidence}")
print(f"Secondary Categories: {result.secondary_categories}")
```

Output:

```
primary_category='Technology' confidence=0.8 secondary_
categories=[{'Account Management': 0.6}, {'Information Technology': 0.4}]

Primary Category: Technology
Confidence: 0.8
Secondary Categories: [{'Account Management': 0.6}, {'Information
Technology': 0.4}]
```

Understanding Structured Output Parsing

1. **Pydantic Models:** Define the exact structure of the expected output, including data types and validation rules.

2. **Format Instructions:** The parser automatically generates instructions that tell the LLM how to format its response to match the expected structure.

3. **Type Safety:** The parsed result is a typed Python object, preventing runtime errors from unexpected response formats.

4. **Rich Information:** Instead of just a category, you get confidence scores and alternative possibilities, enabling more sophisticated decision-making in your application.

Integrating Traditional ML Models

The framework LangChain not only excels with LLM integration, it also works well with traditional ML classification models.

While LLMs are powerful, sometimes you need the speed, cost-efficiency, or specific performance characteristics of traditional machine learning models. LangChain makes it easy to integrate scikit-learn and other ML libraries:

```
from sklearn.feature_extraction.text import TfidfVectorizer
from sklearn.naive_bayes import MultinomialNB
from sklearn.pipeline import Pipeline
import numpy as np

# Training data
texts = ["I love this product", "This doesn't work", "How do I install this?"]
labels = ["positive", "negative", "question"]

# Create a scikit-learn pipeline
ml_classifier = Pipeline([
    ('tfidf', TfidfVectorizer()),
    ('clf', MultinomialNB())
])

ml_classifier.fit(texts, labels)

# Wrap in a LangChain tool
from langchain.tools import Tool

def classify_text(text):
    pred = ml_classifier.predict([text])[0]
    proba = ml_classifier.predict_proba([text])[0]
    confidence = np.max(proba)
    return {"classification": pred, "confidence": float(confidence)}
```

```
classification_tool = Tool(
    name="TextClassifier",
    func=classify_text,
    description="Classifies text as positive, negative, or question"
)

#example usage
text_to_classify = "This is an awesome product!"
result = classification_tool.run(text_to_classify)
print(result)
```

Advantages of Hybrid Approaches

- **Speed:** Traditional ML models typically have lower latency than LLM API calls.
- **Cost:** No per-token charges for inference.
- **Privacy:** Models can run entirely on-premises.
- **Consistency:** Deterministic outputs for the same input.
- **Integration:** LangChain's Tool interface makes it seamless to combine with LLM-based components.

Model Selection Strategies

How do we choose the right model for the NLP task? Below example shows how we can get the best model selected.

In the example below, we are using an additional model from Anthropic to compare performance between models. Please ensure that you add the Anthropic API key for this experiment. Model selection is often overlooked but critical for production success. Different models would excel at different tasks, and the "best" model depends on your specific requirements:

```
from langchain.chains import LLMChain
from langchain.evaluation import load_evaluator
import pandas as pd

def evaluate_models_on_task(models, task_examples, evaluation_criteria):
    results = []
```

CHAPTER 7   LANGCHAIN AND NLP

```python
    # Create evaluator
    evaluator = load_evaluator("labeled_score", criteria=evaluation_criteria)

    for model_name, model in models.items():
        chain = LLMChain(llm=model, prompt=task_examples["prompt"])

        # Run evaluations
        scores = []
        for example in task_examples["examples"]:
            prediction = chain.run(example["input"])
            score = evaluator.evaluate_strings(
                prediction=prediction,
                reference=example["expected"]
            )
            scores.append(score["score"])

        results.append({
            "model": model_name,
            "avg_score": sum(scores) / len(scores),
            "min_score": min(scores),
            "max_score": max(scores)
        })

    return pd.DataFrame(results)

# Example usage
models = {
    "gpt-3.5-turbo": llm,
    "gpt-4": llm,
    "claude-3-sonnet": ChatAnthropic(model="claude-3-sonnet-20240229")
}

classification_examples = {
    "prompt": PromptTemplate(
        template="Classify the sentiment: {text}",
        input_variables=["text"]
    ),
    "examples": [
```

```
        {"input": {"text": "I love this product"}, "expected": "positive"},
        {"input": {"text": "This is terrible"}, "expected": "negative"},
        # Add more examples
    ]
}

results = evaluate_models_on_task(
    models,
    classification_examples,
    "correctness"
)
print(results)
```

Systematic Model Evaluation

This evaluation framework helps you make data-driven decisions about model selection by

1. **Standardized Testing:** Same examples across all models ensure fair comparison.

2. **Multiple Metrics:** Average, minimum, and maximum scores reveal consistency.

3. **Automated Process:** Reduces manual effort and human bias in evaluation.

4. **Extensible Framework:** Easy to add new models or evaluation criteria.

Fine-Tuning for Specific Tasks

LangChain also helps us in fine-tuning for the NLP tasks. When off-the-shelf models don't meet our requirements, fine-tuning can sometimes dramatically improve performance for some specific NLP tasks.

Understanding When to Fine-Tune

Fine-tuning is a powerful technique, but it's not always necessary. The decision matrix below helps you choose the right approach:

CHAPTER 7   LANGCHAIN AND NLP

Decision Factor	Fine-Tuning	Prompting
Task requires consistent formatting	Yes	No
High volume of similar queries	Yes	No
Need for reduced latency	Yes	No
Specialized domain knowledge	Yes	No
Confidential data considerations	Yes	No
General knowledge tasks	No	Yes
Infrequent or varied queries	No	Yes
Rapidly changing requirements	No	Yes

When Fine-Tuning Makes Sense

- **Consistency:** When you need highly consistent output formatting that's difficult to achieve through prompting alone

- **Domain Expertise:** For specialized fields like medical diagnosis, legal analysis, or technical documentation

- **Volume:** When processing thousands of similar requests where fine-tuning costs are offset by reduced per-token charges

- **Latency:** When response speed is critical and a smaller fine-tuned model can outperform larger general models

- **Privacy:** When data sensitivity requires on-premises model deployment

Preparing Fine-Tuning Data with LangChain

LangChain also provides tools to help prepare and format data for fine-tuning. Data preparation is the most time-taking part of fine-tuning. LangChain streamlines this process. This is the dummy code for demonstration:

```
from langchain_community.document_loaders import CSVLoader
from langchain.prompts import FewShotPromptTemplate
import json

# Load example data
```

```python
loader = CSVLoader("customer_support_data.csv")
data = loader.load()

# Format for fine-tuning
def prepare_openai_fine_tuning_data(data, instruction):
    formatted_data = []
    for item in data:
        example = {
            "messages": [
                {"role": "system", "content": instruction},
                {"role": "user", "content": item.page_content},
                {"role": "assistant", "content": item.metadata["response"]}
            ]
        }
        formatted_data.append(example)
    return formatted_data

instruction = "You are a customer support assistant. Provide helpful, accurate, and concise responses."

fine_tuning_data = prepare_openai_fine_tuning_data(data, instruction)

# Save in JSONL format
with open("fine_tuning_data.jsonl", "w") as f:
    for item in fine_tuning_data:
        f.write(json.dumps(item) + "\n")
```

Data Preparation Best Practices

1. **Consistent System Messages:** Use the same system instruction across all training examples.

2. **Quality over Quantity:** 100 high-quality examples often outperform 1000 mediocre ones.

3. **Format Matching:** Ensure training data matches your intended use case format.

4. **JSONL Format:** Required format for OpenAI fine-tuning, with one JSON object per line.

CHAPTER 7 LANGCHAIN AND NLP

Fine-Tuning Language Models

Once you've fine-tuned a model, integrating it into your LangChain has a workflow that is straightforward.

This comprehensive example shows how to fine-tune a GPT-2 model using the Transformers library:

> **Note** For the example, we need hugging face API token. Please create HUGGINGFACEHUB_API_TOKEN using the hugging face account.

```
from transformers import GPT2Tokenizer, GPT2LMHeadModel, Trainer, TrainingArguments, DataCollatorForLanguageModeling
from datasets import load_dataset
import json
import torch

# Load tokenizer and model
model_name = "gpt2"
tokenizer = GPT2Tokenizer.from_pretrained(model_name)
model = GPT2LMHeadModel.from_pretrained(model_name)

tokenizer.pad_token = tokenizer.eos_token
model.config.pad_token_id = model.config.eos_token_id

# Load the dataset
data_files = {"train": "/content/fine_tuning_data.jsonl"}
dataset = load_dataset("json", data_files=data_files)

# Prepare prompt-response pairs
def process_messages(example):
    system_prompt = ""
    user_input = ""
    assistant_response = ""

    for message in example["messages"]:
        if message["role"] == "system":
            system_prompt = message["content"]
        elif message["role"] == "user":
```

```python
            user_input = message["content"]
        elif message["role"] == "assistant":
            assistant_response = message["content"]

    # Combine into a single training sample
    full_prompt = f"System: {system_prompt}\nUser: {user_input}\
nAssistant: "
    full_completion = assistant_response

    return {
        "input_text": full_prompt,
        "output_text": full_completion
    }

# Apply processing
processed_dataset = dataset.map(process_messages, remove_
columns=["messages"])

# Tokenize
def tokenize_function(examples):
    inputs = [i + o for i, o in zip(examples["input_text"],
    examples["output_text"])]
    model_inputs = tokenizer(inputs, truncation=True, padding="max_length",
    max_length=512)
    return model_inputs

tokenized_dataset = processed_dataset.map(tokenize_function, batched=True)

# Data collator
data_collator = DataCollatorForLanguageModeling(tokenizer=tokenizer,
mlm=False)

# Training arguments
training_args = TrainingArguments(
    output_dir="./gpt2-finetuned",
    overwrite_output_dir=True,
    num_train_epochs=3,
    per_device_train_batch_size=2,
    save_steps=500,
```

```
    save_total_limit=2,
    logging_steps=100,
    # evaluation_strategy="no",
    weight_decay=0.01,
    push_to_hub=False,
    fp16=torch.cuda.is_available(),
)

# Trainer
trainer = Trainer(
    model=model,
    args=training_args,
    train_dataset=tokenized_dataset["train"],
    data_collator=data_collator,
)

# Train
trainer.train()

# Save
model.save_pretrained("./gpt2-finetuned")
tokenizer.save_pretrained("./gpt2-finetuned")
```

Fine-Tuning Process Breakdown

1. **Model Loading:** Start with a pre-trained model (GPT-2 in this case) as the foundation.

2. **Data Processing:** Convert conversation format into training sequences.

3. **Tokenization:** Convert text into numerical tokens the model can process.

4. **Training Configuration:** Set hyperparameters like learning rate, batch size, and number of epochs.

5. **Training Loop:** The Trainer class handles the complex training process.

6. **Model Saving:** Persist the fine-tuned model for later use.

Using the Fine-Tuned Model for Inference

After fine-tuning, you can use your specialized model for inference:

```
from transformers import pipeline, GPT2Tokenizer, GPT2LMHeadModel

# 1. Load the fine-tuned GPT-2 model
model_path = "./gpt2-finetuned"  # path where your fine-tuned model is saved
tokenizer = GPT2Tokenizer.from_pretrained(model_path)
model = GPT2LMHeadModel.from_pretrained(model_path)

# Create a pipeline for text generation
generator = pipeline("text-generation", model=model, tokenizer=tokenizer, max_length=512)

# 2. Define the function to handle text generation
def generate_response(input_text):
    # Generate the response using the model pipeline
    result = generator(input_text, return_full_text=False)
    return result[0]['generated_text']

# 3. Example usage
query = "I can't reset my password"
response = generate_response(query)
print("\nResponse from fine-tuned model:\n")
print(response)
```

Production Deployment Considerations

1. **Model Versioning:** Keep track of different model versions and their performance.

2. **A/B Testing:** Compare fine-tuned models against baseline models with real traffic.

3. **Monitoring:** Track model performance metrics in production.

CHAPTER 7　LANGCHAIN AND NLP

4. **Fallback Strategies:** Have backup models ready if the fine-tuned model fails.

5. **Resource Management:** Fine-tuned models may have different memory and compute requirements.

Integration with LangChain Workflows

After we have a fine-tuned model, we can integrate it seamlessly into workflows by creating custom LLM wrappers or using LangChain's existing integrations with popular model serving platforms.

We can easily switch between different models for different tasks within the same application, or gradually migrate to specialized ones the use case evolves.

Key Takeaways

- LangChain enables seamless integration of NLP capabilities into workflows.

- You can build and customize classifiers tailored to specific use cases.

- Model selection and tuning provide flexibility to optimize performance for different tasks.

In this chapter, we explored how LangChain empowers you to integrate NLP capabilities from building custom classifiers to selecting and tuning models for specific needs; you now have a solid foundation to apply NLP techniques within LangChain workflows.

As we transition to the next chapter, we shift focus from understanding text to taking intelligent actions. Get ready to explore how to build AI agents using LangGraph, where you'll learn to orchestrate decision-making, tool use, and dynamic reasoning in multi-step workflows.

CHAPTER 8

Building AI Agents with LangGraph

An AI **agent** is an entity that uses an LLM to perform complex, goal-oriented tasks by dynamically deciding what actions to take next. In the context of LangChain, agents operate using a loop in which they

1. **Receive an input or goal**
2. **Reason** about the next step using an LLM
3. **Select a tool or action** (e.g., search, calculator, API call)
4. **Execute the action**
5. **Observe the result**
6. **Repeat the process** until the goal is met or a stopping condition is reached

Agents rely on **tool usage**, **planning**, and sometimes **memory** to operate effectively. For instance, a research agent might use a web search tool to gather data, a summarizer to condense it, and a planner to decide how to present the information.

Agents allow LLMs to move from one-shot prompting to **multi-step reasoning**, enabling applications like AI assistants, task automators, and decision support systems. In LangChain, the agent framework abstracts these behaviors, making it easier to orchestrate complex workflows powered by LLMs.

LLM vs. Agents

LLMs are fundamentally **stateless** and **reactive**: they generate responses based solely on the given input and their learned parameters, without memory of past interactions or the ability to take autonomous decisions over time.

In contrast, **agents** introduce a layer of **autonomy and reasoning** on top of LLMs. Agents use LLMs as a reasoning engine but are guided by additional frameworks that allow them to make decisions, use tools, store and retrieve memory, and pursue goals over multiple steps. While an LLM can generate a helpful response to a single question, an agent can **break down a task**, **choose the right tools**, **interact with an environment**, and **manage its own state** to achieve complex objectives.

Thus, the distinction lies in functionality: LLMs provide raw intelligence, while agents provide structured decision-making capabilities built around that intelligence.

Core Components of LangGraph

Graphs

The fundamental building block in LangGraph is the graph itself. Graphs define the overall structure and flow of your application:

```
from langgraph.graph import StateGraph
# Create a graph with a specific state type
graph = StateGraph(StateType)
```

Graphs contain

- Nodes (components that perform operations)
- Edges (connections between nodes that direct the flow)
- State (information that persists throughout execution)

States and State Management

State management is critical in LangGraph. States allow information to persist across steps of execution:

```
from typing import TypedDict, Annotated
# Define a typed state
class AgentState(TypedDict):
    messages: Annotated[list[BaseMessage], operator.add]
    memory: dict
```

CHAPTER 8   BUILDING AI AGENTS WITH LANGGRAPH

States can include

- Conversation history
- Tool outputs and intermediate calculations
- Memory systems for long-term context
- Agent-specific knowledge

Nodes

Nodes are the functional components that perform operations within your graph:

```
# Add nodes to the graph
graph.add_node("agent", agent_node)
graph.add_node("tool_executor", tool_executor)
```

Common node types include

- LLM-powered agents that generate text responses
- Tool executors that perform specific tasks
- Memory components that store and retrieve information
- Routers that direct the flow based on conditions

Edges and Conditional Routing

Edges define how information flows between nodes:

```
# Add basic edge
graph.add_edge("agent", "tool_executor")

# Add conditional edge
graph.add_conditional_edges(
    "agent",
    condition_function,
```

```
    {
        "use_tool": "tool_executor",
        "respond": "output"
    }
)
```

Edges can be

- Simple (always passing from one node to another)
- Conditional (routing based on output of a condition function)
- Dynamic (changing based on the state or other factors)

Basic Structure of a LangGraph Agent

A typical LangGraph Agent follows this pattern which is mentioned in the psuedo code below:

1. Define the state structure.
2. Create nodes for different components.
3. Connect nodes with edges to form a graph.
4. Execute the graph with initial inputs.

```
# 1. Define state
class ConversationState(TypedDict):
    messages: list
    current_question: str

# 2. Create nodes
def agent_node(state: ConversationState):
    # Agent logic here
    return {"messages": updated_messages}
```

```
# 3. Build graph
graph = StateGraph(ConversationState)
graph.add_node("agent", agent_node)
graph.add_edge("agent", "output")
graph.set_entry_point("agent")

# 4. Compile and run
app = graph.compile()
result = app.invoke({"messages": [], "current_question": "Tell me about LangGraph"})
```

Creating Autonomous Agents Using LangGraph

Autonomous agents represent an advanced application of LLMs where the AI system can independently pursue goals through reasoning, planning, and action. LangGraph provides an excellent framework for building these agents due to its stateful, graph-based approach. Let's explore how to create autonomous agents using LangGraph.

An autonomous agent in LangGraph typically consists of

1. **Goal-Directed Behavior:** The agent works toward specific objectives.
2. **Reasoning Capabilities:** The agent can plan and adjust its approach.
3. **Tool Usage:** The agent can interact with external systems.
4. **Memory:** The agent maintains context across steps.
5. **Action Loop:** The agent follows a cycle of thought and action.

The most common pattern for implementing these agents is the ReAct (Reasoning + Acting) pattern, which involves

- Thinking about what to do (Reasoning)
- Performing actions (Acting)
- Observing results
- Updating the approach

CHAPTER 8   BUILDING AI AGENTS WITH LANGGRAPH

Defining Agent State for Autonomous Agents

Agent state serves as the memory and context container for your autonomous agent. It's a structured representation of everything the agent knows and has experienced during its execution. When building autonomous agents with LangGraph, clearly defining this state structure is a crucial first step.

Why State Matters for Autonomous Agents?

In traditional conversational AI applications, each interaction is often handled independently. Autonomous agents, however, need persistent memory to function effectively. The state structure allows your agent to

- Remember previous interactions and their outcomes
- Track progress toward goals
- Maintain awareness of available tools and their prior usage
- Store intermediate results from multi-step reasoning
- Build and update mental models of the task environment

Components of a Well-Designed Agent State

A comprehensive agent state typically includes several key elements:

1. **Message History:** The conversation transcript between the user and agent, providing context for decision-making
2. **Goal Tracking:** The current objective or task the agent is working to accomplish
3. **Tool Usage Records:** Documentation of which tools have been used, when, and with what results
4. **Intermediate Steps:** The sequence of reasoning steps and actions the agent has taken
5. **Memory Store:** A flexible container for additional information the agent might need to remember

The Agent State serves as the memory of your system, tracking all relevant data that flows through your agent (like inputs, intermediate results, and final outputs). Agent Components are the building blocks that handle specific tasks, typically including language models, prompts, and other tools that transform or process information.

Agent Nodes: Each node represents a distinct stage in the workflow, where a component processes the current state and generates a new state, forming the vertices of the graph.

Routing Logic: Governs the navigation through the graph by determining the next node to activate based on the current state or deciding when to conclude execution, effectively establishing the edges connecting nodes.

Graph Construction: Involves assembling the entire structure by defining nodes, their connections, entry points, and conditional pathways, creating the agent's decision-making framework.

Execution initiates the process by providing initial state values, allowing the agent to traverse the graph according to its defined logic until reaching a terminal condition. Together, these elements enable LangGraph to create sophisticated, multi-step reasoning processes with clear state management and control flow.

```python
from langchain_core.prompts import ChatPromptTemplate
from langchain_openai import ChatOpenAI
from langgraph.graph import StateGraph, END

# 1. Define Agent State
class AgentState(dict):
    text: str

# 2. Define Agent Components
llm = ChatOpenAI(model="gpt-3.5-turbo")
prompt = ChatPromptTemplate.from_template("Answer this question: {text}")
chain = prompt | llm

# 3. Create Agent Nodes
def process(state):
    response = chain.invoke({"text": state["text"]})
    state["response"] = response.content
    return state

# 4. Define Routing Logic
def should_continue(state):
    if "response" in state:
        return END
    return "process"
```

```
# 5. Build the Graph
graph = StateGraph(AgentState)
graph.add_node("process", process)
graph.set_entry_point("process")
graph.add_conditional_edges("process", should_continue)
agent = graph.compile()

# 6. Run the Agent
result = agent.invoke({"text": "What is the capital of France?"})
print(result["response"])
```

Agent Architectures

Agent architectures in LangGraph define the structural organization and decision-making patterns that determine how agents process information, make decisions, and interact with their environment. The right architecture choice significantly impacts an agent's capabilities, efficiency, and suitability for specific tasks.

ReAct Architecture

The ReAct (Reasoning and Acting) pattern represents one of the most fundamental agent architectures in LangGraph. This architecture interleaves reasoning and action taking in a structured manner:

- **Observation:** The agent receives information from its environment.
- **Thought:** The agent reasons about the observation and its current state.
- **Action:** The agent selects an appropriate action to take.
- **Result:** The agent observes the outcome of its action.

Reflexion Architecture

The Reflexion architecture extends ReAct by adding a critical self-reflection component:

- Execute a ReAct-style reasoning and action loop.

- Reflect on the outcome and strategy used.
- Refine approach based on reflection.

This architecture enables agents to learn from their experiences and improve over time.

Plan-and-Execute Architecture

The Plan-and-Execute architecture separates planning from execution:

- **Planner:** Generates a detailed, multi-step plan
- **Executor:** Carries out individual steps in the plan
- **Monitor:** Tracks progress and determines when **preplanning** is needed

Multi-agent Architectures

LangGraph allows constructing sophisticated systems with multiple cooperating agents:

- **Heterogeneous Teams:** Different agents with specialized capabilities
- **Collaborative Problem-Solving:** Agents that coordinate to solve tasks
- **Market-Based Systems:** Agents that compete or bid for tasks

Case Study
Reflection-Based Agents for Content Generation

This case study illustrates the use of agentic workflows to iteratively refine content generation using LangGraph. The approach leverages a modular state machine framework where each state (or node) corresponds to a distinct task in the content development pipeline. The goal is to generate high-quality written content based on a given topic, with the system employing a reflection mechanism to improve successive drafts before finalization.

The agent is structured around three core functional nodes: content generation, reflection, and finalization. Initially, the "generate" node produces a first draft based solely on the input topic. In subsequent iterations, the agent uses feedback from the "reflect" node to improve the draft. This reflection is generated by prompting the LLM to critically evaluate the quality of the previous draft with respect to clarity, relevance, engagement, and completeness. This feedback is then used in the next round of generation to improve the draft. A conditional logic function, should_continue, governs whether the system should continue refining the draft or proceed to finalization, based on a pre-set maximum number of iterations.

LangGraph's StateGraph is used to implement this agent loop, with transitions managed via explicit edges and conditional logic. The state itself is a TypedDict named AgentState, which maintains not only the messages exchanged and the current draft but also the number of iterations and the reflection content. This allows the agent to retain memory across iterations, simulating an intentional and iterative authoring process. The final output is a polished draft that has been iteratively improved based on internal critiques generated by the model itself.

This reflective agent demonstrates how agentic patterns can introduce self-corrective behavior in generative pipelines. The architecture is particularly valuable for scenarios that demand refinement and quality control, such as educational content creation, professional writing, or marketing copy generation. By encapsulating both generation and reflection in a graph-based workflow, the agent offers modularity, interpretability, and scalabilitykey traits for real-world adoption.

Agentic Text-to-SQL Generator Using LangGraph

The second case study presents an intelligent agent that translates natural language queries into executable SQL commands using a LangGraph workflow. This setup showcases the powerful synergy between prompt engineering, database access, and agentic reasoning. The use case is framed around a hypothetical employee database hosted in SQLite, allowing the agent to parse user intent and fetch relevant data programmatically.

The architecture follows a structured agentic design, integrating multiple tools and decision points. It begins by defining tools (in LangChain's parlance) that can validate and execute SQL queries. The main task is decomposed into interpreting user input, generating candidate SQL queries using the LLM, validating them, executing them against a live database, and optionally retrying in case of failure. Each stage is modeled as a node in a LangGraph, and transitions are determined based on intermediate results, such as whether the generated SQL is syntactically and semantically valid.

Technically, the agent maintains a state dictionary that stores the user input, generated SQL, results, and error messages. This persistent state across steps enables robust error handling and retry mechanisms—key features in real-world applications. The workflow uses LangGraph's StateGraph class to formalize transitions between stages like generation, validation, execution, and output. The SQL generation prompt is carefully templated to maximize the chance of producing accurate SQL commands based on user-provided intent, leveraging LangChain's ChatPromptTemplate and StrOutputParser.

This case study exemplifies how agentic systems can be used to construct sophisticated, multi-step LLM workflows with real-world utility. It serves as a practical blueprint for building natural language interfaces over databases, which is a growing area of interest in enterprise and analytics platforms. By isolating concerns into nodes and maintaining a shared state, LangGraph enables extensibility and maintainability while preserving the context across task boundaries.

In this chapter, we explored how LangGraph enables the construction of powerful, autonomous AI agents. By understanding its core components and architecture, you've learned how to design agents capable of making decisions.

Key Takeaways

- LangGraph supports graph-based, stateful agent workflows.
- Reflection and memory integration improve agent reasoning.
- LangGraph makes agents more modular, transparent, and maintainable.

As we move forward, the next chapter focuses on integrating LangChain with external tools and frameworks. You'll learn how to enhance your agents by connecting them to real-time APIs, leveraging the capabilities of machine learning libraries like TensorFlow and PyTorch, and creating hybrid AI systems that unlock the full power of the LangChain ecosystem.

CHAPTER 9

LangChain Framework Integration

LangChain is powerful on its own, but its true potential shines when integrated with other AI tools and frameworks. In this chapter, we explore how to connect LangChain to external APIs, combine it with popular machine learning libraries like TensorFlow and PyTorch, and build hybrid AI architectures.

Working with External APIs

External APIs allow LangChain agents and chains to interact with the world beyond the local environment. APIs can fetch real-time data, perform specialized computations, or tap into third-party services like databases, customer support systems, or weather information.

API Integration Best Practices

Before integrating APIs with LangChain, it is essential to follow certain best practices to ensure reliability, security, and performance:

- **Authentication Handling:** Securely store and manage API keys using environment variables or secrets managers.

- **Error Handling and Retries:** Always account for potential failures with appropriate error handling and retry logic.

- **Rate Limiting Awareness:** Respect the rate limits set by APIs to avoid service disruptions.

CHAPTER 9   LANGCHAIN FRAMEWORK INTEGRATION

- **Input and Output Validation:** Validate inputs before sending API requests and validate responses to prevent downstream errors.
- **Asynchronous Execution:** Use async calls for better performance, especially when dealing with slow or rate-limited APIs.

> **Tip** LangChain natively supports both synchronous and asynchronous executions for API-based tools.

Using LangChain with Popular APIs

Let's walk through an example of using LangChain with OpenWeatherMap API to fetch weather data; refer to Figure 9-1 for the flow.

Step 1: Create a Custom Tool

```python
from langchain.tools import BaseTool
import requests

class WeatherTool(BaseTool):
    name: str = "weather_tool"
    description: str = "Fetches weather information for a given city."

# <-- make sure to define your API key properly
    def _run(self, city: str) -> str:
        api_key = openWeatherAPI_key

url = f"http://api.openweathermap.org/data/2.5/weather?q={city}&appid={api_key}&units=metric"
        response = requests.get(url)
        data = response.json()
        if response.status_code == 200:
            return f"The weather in {city} is {data['weather'][0]['description']} with a temperature of {data['main']['temp']} °C."
        else:
```

```
            return f"Failed to fetch weather data: {data.get('message',
            'Unknown error')}"

    async def _arun(self, city: str) -> str:
        raise NotImplementedError("Async not implemented yet.")
weather_tool = WeatherTool()
city = "London"
result = weather_tool._run(city)
print(result)
```

Output:
The weather in London is overcast clouds with a temperature of 7.75 °C.

The WeatherTool class extends LangChain's BaseTool and defines a _run method that fetches weather data from the OpenWeather API for a given city. The method constructs the API request URL using the provided city and an API key and then sends a request to the OpenWeather API. If the request is successful (status code 200), it parses the JSON response and returns a formatted string with the weather description and temperature in Celsius. If the request fails, it returns an error message. The class also includes an unimplemented async version (_arun) for future use.

Step 2: Use the Tool in an Agent

```
from langchain.agents import initialize_agent, AgentType
from langchain.llms import OpenAI

weather_tool = WeatherTool()
agent = initialize_agent(
    tools=[weather_tool],
    llm=llm,
    agent=AgentType.ZERO_SHOT_REACT_DESCRIPTION)

response = agent.run("What's the weather like in Dublin today?")
print(response)
```

Output:
I am unable to answer the original question as I am unable to fetch the weather for Dublin.

CHAPTER 9   LANGCHAIN FRAMEWORK INTEGRATION

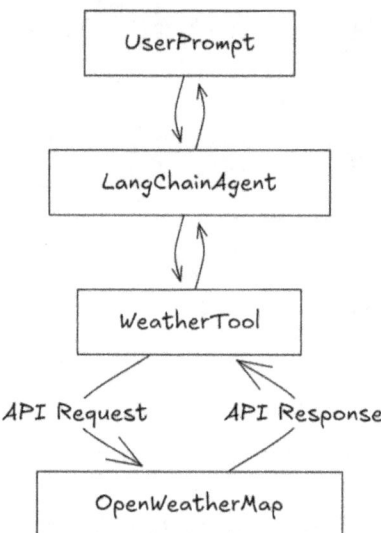

Figure 9-1. *LangChain agent fetching weather data*

The code integrates the WeatherTool into a LangChain agent by initializing it with the initialize_agent function, passing the tool and an OpenAI LLM as inputs. The agent is set up to use a zero-shot reaction strategy (AgentType.ZERO_SHOT_REACT_ DESCRIPTION), meaning it responds to queries without prior training specific to the task. However, in this case, the agent is unable to fetch weather data for Dublin, and it responds with the message: *"I am unable to answer the original question as I am unable to fetch the weather for Dublin."* This suggests that either the tool isn't properly connected to an API or an issue occurred in fetching the data.

Combining LangChain with Other Frameworks

Beyond APIs, LangChain can be deeply integrated with machine learning libraries like TensorFlow, PyTorch, and Hugging Face Transformers to build even more powerful applications.

TensorFlow, PyTorch, and More

LangChain can utilize the predictive capabilities of deep learning models. Here's how you can integrate a custom PyTorch model into a LangChain tool.

Example: Sentiment Analysis with a PyTorch Model

First, let's assume you have a fine-tuned sentiment analysis model in PyTorch.

```python
import torch
from transformers import BertTokenizer, BertForSequenceClassification
from langchain.tools import BaseTool
from typing import Optional
from pydantic import Field, model_validator

class SentimentAnalysisTool(BaseTool):
    name: str = "sentiment_analysis"
    description: str = "Classifies the sentiment of a given text."
    model_name: Optional[str] = Field
    (default="nlptown/bert-base-
    multilingual-uncased-sentiment")

    # These are not Pydantic fields, they will be set after initialization
    tokenizer: BertTokenizer = None
    model: BertForSequenceClassification = None

    @model_validator(mode="after")
    def load_model(self) -> "SentimentAnalysisTool":
        self.tokenizer = BertTokenizer.from_pretrained(self.model_name)
        self.model = BertForSequenceClassification.from_pretrained(self.
        model_name)
        return self

    def _run(self, text: str) -> str:
        inputs = self.tokenizer(text, return_tensors="pt", truncation=True,
        padding=True)
        outputs = self.model(**inputs)
        probs = torch.nn.functional.softmax(outputs.logits, dim=1)
        sentiment_class = torch.argmax(probs, dim=1).item()

        if sentiment_class <= 1:
            return "Negative"
        elif sentiment_class == 2:
            return "Neutral"
```

```
        else:
            return "Positive"

    async def _arun(self, text: str) -> str:
        raise NotImplementedError("Async not implemented yet.")

# Instantiate the tool
tool = SentimentAnalysisTool()

# Run sentiment analysis
text = "I love this product, it's amazing!"
result = tool._run(text)

print(result)
```

Output:
Positive

The SentimentAnalysisTool class is a custom LangChain tool for performing sentiment analysis using a pre-trained BERT model. It loads the BERT tokenizer and model (nlptown/bert-base-multilingual-uncased-sentiment) during initialization via the model_validator. The _run method processes the input text by tokenizing it, passing it through the model, and applying softmax to get the probabilities of each sentiment class. Based on the class with the highest probability, it returns either "Negative," "Neutral," or "Positive" sentiment. The tool is then instantiated, and a sample text ("I love this product, it's amazing!") is analyzed, outputting the sentiment (in this case, "Positive").

Now, this custom tool can be embedded into an agent, chain, or workflow just like any native LangChain tool.

Pro Tip You can use TensorFlow models similarly by wrapping prediction functions into a custom tool.

Hybrid AI Architectures

Hybrid architectures combine LLM reasoning with symbolic reasoning, custom models, and traditional ML pipelines. LangChain provides the scaffolding to orchestrate such complex workflows. Refer to the flow in the Figure 9-2.

CHAPTER 9   LANGCHAIN FRAMEWORK INTEGRATION

- **Step 1:** LangChain agent parses a user query.
- **Step 2:** If the task involves text classification, route to a custom PyTorch model.
- **Step 3:** Otherwise, use the LLM for reasoning.
- **Step 4:** Aggregate results and respond.

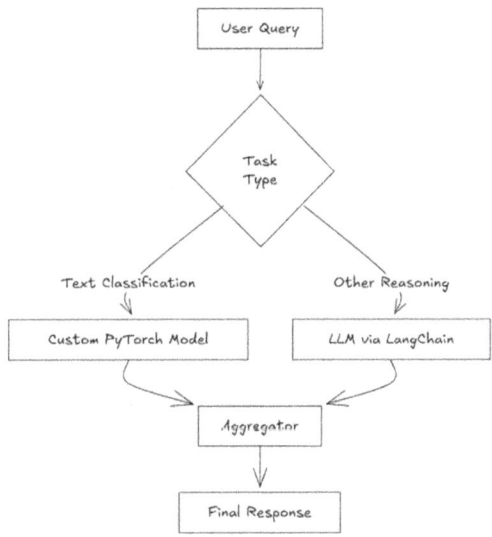

Figure 9-2. *Hybrid AI Flow*

Router Example Code

```
# Imports
from langchain.llms import HuggingFaceHub, AzureOpenAI from langchain.
agents import AgentExecutor, create_openai_functions_agent from langchain.
tools import Tool from langchain_core.prompts import ChatPromptTemplate,
MessagesPlaceholder from langchain.schema import SystemMessage from
langchain.tools import BaseTool from pydantic import BaseModel, Field from
typing import Optional, Type

# Reuse components from your uploaded file
# llm is already created: (AzureOpenAI or HuggingFaceHub)

# Reuse SentimentAnalysisTool
# Make sure you have this class already imported from your code:
# class SentimentAnalysisTool(BaseTool): ...
```

```python
sentiment_tool = SentimentAnalysisTool()

# Create a lightweight LLM reasoning function
def llm_reasoning_function(query: str) -> str:
    """Handles general reasoning queries via LLM."""
    return llm.invoke(query)

# Router function
def router(query: str) -> str:
    """
    Decide whether to use SentimentAnalysisTool or LLM based on task.
    If query mentions 'sentiment' or 'classify', route to
    SentimentAnalysisTool.
    Else, route to LLM.
    """
    keywords = ["sentiment", "classify", "classification"]
    if any(keyword in query.lower() for keyword in keywords):
        return "sentiment_analysis"
    else:
        return "llm_reasoning"

# Set up tools
tools = [
    Tool.from_function(
        name="sentiment_analysis",
        description="Classifies text sentiment using a custom PyTorch
        model.",
        func=sentiment_tool._run,
    ),
    Tool.from_function(
        name="llm_reasoning",
        description="Handles general reasoning queries with an LLM.",
        func=llm_reasoning_function,
    )
]
```

```python
# Custom agent prompt
prompt = ChatPromptTemplate.from_messages([
    SystemMessage(content="You are a hybrid agent that routes tasks either
    to a sentiment classifier or an LLM."),
    MessagesPlaceholder(variable_name="chat_history"),
    MessagesPlaceholder(variable_name="agent_scratchpad"),
])

# Create an agent
agent = create_openai_functions_agent(llm=llm, tools=tools, prompt=prompt)

# Create an executor
agent_executor = AgentExecutor(agent=agent, tools=tools, verbose=True)

# Hybrid Execution function
def hybrid_agent_executor(user_query: str):
    selected_tool_name = router(user_query)
    for tool in tools:
        if tool.name == selected_tool_name:
            response = tool.invoke(user_query)
            break
    return response

# Example Usage
query1 = "Can you classify the sentiment of this text: 'I hate rainy
         days.'?"
query2 = "What's the capital of France?"
query3 = "What's the weather of France today?"

print("Query 1 Result:", hybrid_agent_executor(query1))
print("Query 2 Result:", hybrid_agent_executor(query2))
```

Output:
Query 1 Result: Negative
Query 2 Result: The capital of France is Paris.

This code creates a hybrid agent that routes user queries to either a sentiment analysis tool or a general LLM based on the content of the query. It uses a router function to check for keywords like "sentiment" or "classify" and directs sentiment-related queries to the custom SentimentAnalysisTool, while other queries are handled by the LLM. The agent is set up using LangChain's AgentExecutor, and it efficiently processes queries by invoking the appropriate tool, such as classifying sentiment or providing general knowledge, with examples showing it can classify text sentiment as "Negative" or respond to general queries like the capital of France.

Key Takeaways

- LangChain excels at integrating external APIs securely and effectively.

- You can leverage TensorFlow, PyTorch, and other ML frameworks through custom tools.

- Hybrid architectures empower agents to combine reasoning and predictive intelligence flexibly.

- Modularity in LangChain makes orchestration between LLMs, APIs, and ML models seamless.

The next chapter brings everything together with production-level considerations: deployment, testing, optimization.

CHAPTER 10

Deploying LangChain Applications

Once we have built the LangChain applications, taking it from the proof of concept to production is a pivotal step in the development life cycle. Success of the use case will depend on the well-executed deployment strategy that ensures the application is robust, scalable, and reliable. We can look at some of the important aspects that are important in deploying in real-world environments including architecture decisions, scaling strategies, performance optimizations, testing methodologies, and monitoring practices.

Preparing for Production

Before the application goes live, careful preparation is essential to avoid runtime surprises and ensure a seamless deployment. The following best practices will help you lay a strong foundation for production readiness.

Environment Isolation: Maintain separate environments for development, staging, and production. This minimizes the risk of disrupting live services due to code changes or misconfigurations.

Figure 10-1 illustrates a typical environment isolation setup. Development, staging, and production environments are separated to prevent accidental changes in live systems. The flow shows how code moves from version control through CI/CD pipelines and automated testing into the development environment and only then promoted to staging and finally production. This separation ensures stability, testing integrity, and controlled deployment.

Secrets manager securely handles sensitive data—like API keys and access tokens—using tools such as AWS Secrets manager or Azure key vault.

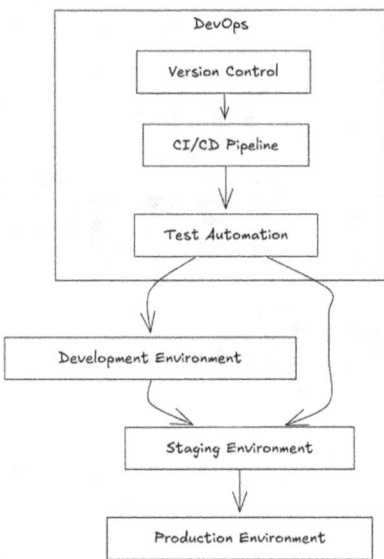

Figure 10-1. Typical environment isolation setup

Configurable modular code separates settings into environment-specific files. This enables flexibility in managing API endpoints, keys, and other deployment-specific parameters.

Comprehensive end-to-end tests ensure your application behaves as expected before deployment.

Input data validation will ensure that we have always sanitized and validated user inputs to prevent unexpected behavior, security risks, or application crashes.

Logging and monitoring provide visibility into performance, anomaly detection, and error tracking. Traceability and logging tools support effective monitoring.

Architecture

Selecting the right deployment architecture is critical for maintaining performance, managing complexity, and scaling with demand. Below are some commonly used approaches.

Monolithic Deployment: In a monolithic architecture, the complete application is bundled and deployed as a single, unified service. This straightforward approach suits small-scale applications or initial prototypes. However, as the application expands, managing and scaling the monolith can become increasingly complex.

Microservices Deployment: A microservices architecture breaks down the LangChain system into smaller, autonomous services, such as distinct services for chains, tools, and agents. This design enhances scalability and adaptability. Tools like Kubernetes facilitate deployment orchestration, service discovery, and maintain high availability.

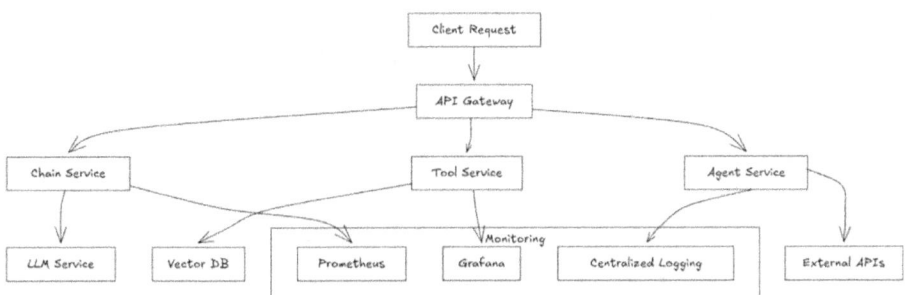

Figure 10-2. Microservices-based architecture for LangChain

Referring to the architecture in Figure 10-2, incoming requests are routed via an API Gateway to specialized services for chains, tools, and agents. Each service interacts with necessary components like LLMs, vector databases, or external APIs. Observability is integrated using Prometheus, Grafana, and centralized logging to monitor the health and performance of each microservice independently.

Serverless Deployment: When considering the stack for applications that are event-driven or experience variable traffic, serverless platforms like AWS Lambda or Azure Functions offer automated scaling and resource optimization. This model is ideal for applications with sporadic workloads, though developers should be mindful of latency caused by cold starts.

Scaling Applications

Scaling applications involves ensuring they remain responsive as load increases. Here's how to approach scaling effectively.

Horizontal Scaling: Deploy additional service instances and employ load balancers to evenly distribute incoming traffic. Implement auto-scaling policies on cloud platforms to automatically adjust resource allocation based on metrics such as CPU or memory utilization.

Vertical Scaling: Increase the computational resources (CPU, RAM) of existing service instances. This can be effective for predictable workloads, but it offers limited scalability compared to horizontal scaling.

Load Balancing

As we can see in Figure 10-3, we can use load balancer that distributes incoming user requests across multiple service instances, each of which can access a shared Redis cache. This architecture can support high availability and scalability, ensuring consistent performance under varying loads while reducing redundant computations via caching.

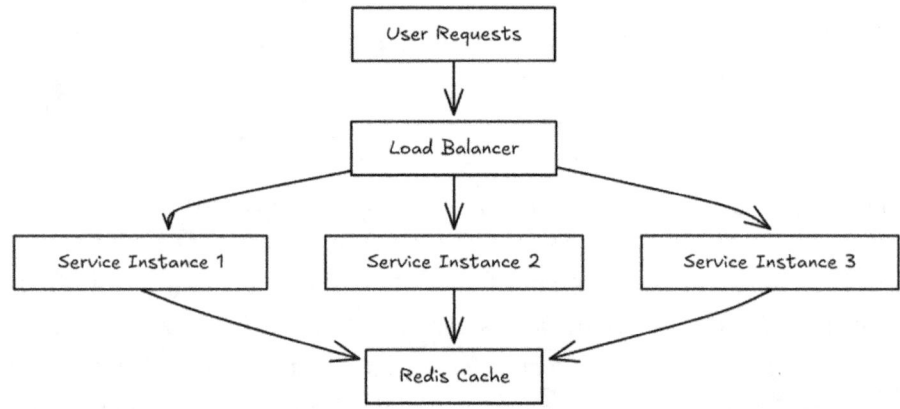

Figure 10-3. *Horizontal scaling in LangChain applications*

Optimizing for Production

Optimizing for production helps improve efficiency, reduce latency, and manage costs. Key areas to focus on include the following.

Caching Results: Cache frequent responses using tools like Redis to reduce backend processing and external API calls. Be sure to implement cache expiration to avoid serving stale data.

From the pipeline shown in Figure 10-4, after validating inputs, it will check for cached results to reduce unnecessary computation. On a cache miss, it proceeds to call the LLM or external API, optimizing the prompt and generating a response. Responses are returned and optionally cached for future use making the system efficient and cost-effective.

CHAPTER 10 DEPLOYING LANGCHAIN APPLICATIONS

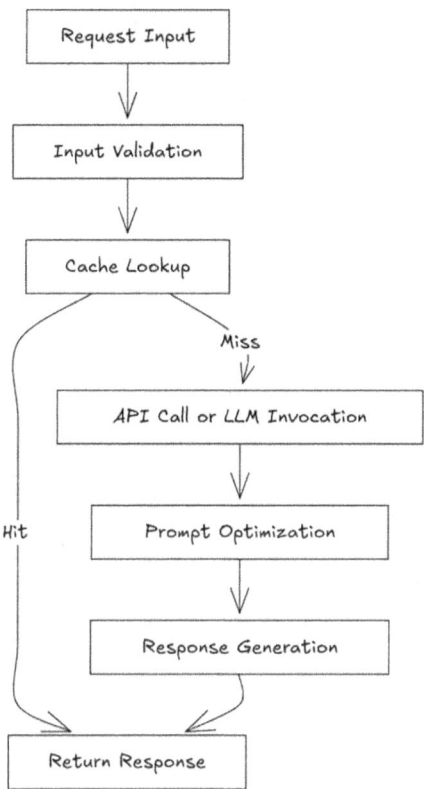

Figure 10-4. *Internal flow of an optimized request pipeline*

Optimizing API Calls: Batching API requests, using pagination, and limiting tokens can significantly improve performance when interacting with external services.

Model Optimization: Model optimization involves selecting appropriate model sizes, using smaller and faster models for non-critical tasks. Design prompts thoughtfully to reduce token consumption, and leverage few-shot learning for more efficient model performance.

Testing and Evaluating Applications

Thorough testing is vital for ensuring stability and quality before deploying applications into production.

The approach shown in Figure 10-5 begins with unit tests for individual components, followed by integration testing to verify how modules work together. In a staging environment, load and stress testing simulate production traffic. Performance evaluation ensures responsiveness and scalability, and A/B testing helps select the best model or configuration before final deployment.

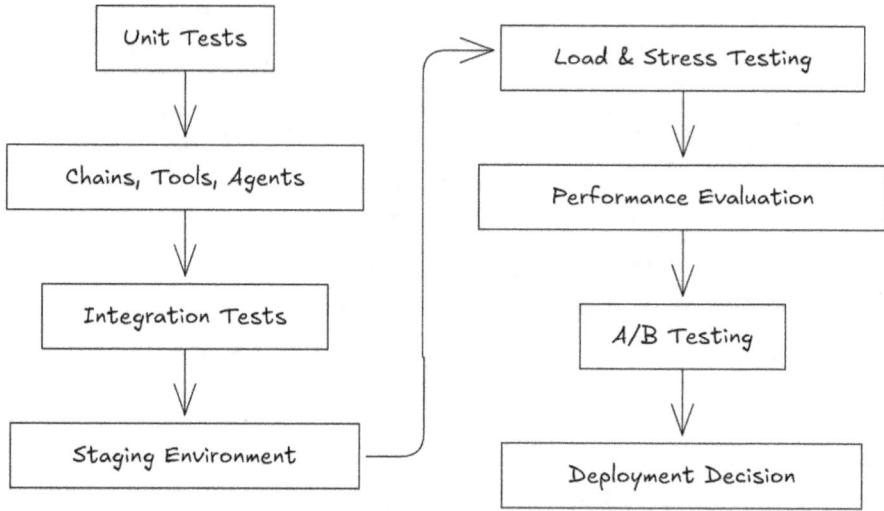

Figure 10-5. Layered approach to testing applications

Monitoring and Logging

Once deployed, maintaining visibility into your application's health is key to proactive issue resolution and continuous improvement.

Real-Time Monitoring: Use tools like Prometheus and Grafana to monitor key metrics—latency, error rates, CPU/memory usage.

Implement structured logging with tools like the ELK stack or AWS CloudWatch to enable centralized log analysis. Include contextual metadata such as request IDs for easier debugging.

Error Handling and Alerts: Set up automated alerting for critical failures using platforms like Sentry or PagerDuty. Provide user-friendly fallback options and error messages in the event of service interruptions.

Key Takeaways

- Taking LangChain applications from proof of concept to production requires a robust and scalable deployment strategy.

- Architecture choices, scaling, and performance optimizations are key for real-world success.

- Rigorous testing and continuous monitoring sustain reliability and long-term performance.

The next chapter will focus on best practices, ethics, and preparing for the evolving landscape of LangChain and AI agents.

CHAPTER 11

Best Practices and Practical Aspects

This concluding chapter of the book focuses on building ethical and scalable AI systems using LangChain. We will explore key principles for responsible AI, learn to optimize applications, avoid common pitfalls, and stay informed about emerging trends.

Building Ethical and Compliant AI Systems
Ethical Considerations in AI Development

As AI becomes an integral part of our applications, building ethically responsible systems is no longer optional. In LangChain-based solutions, ethical considerations are crucial when designing prompts, agents, and decision workflows.

Key Principles

- **Fairness:** Ensure models do not inadvertently perpetuate biases. Train on diverse datasets and carefully validate outputs for fairness across different user groups.

- **Transparency:** Make the decision logic behind agents understandable and explainable. Implement explainability features that show users how a recommendation or decision was made.

- **Accountability:** Maintain logs of agent actions and provide human-in-the-loop mechanisms where possible. Set up governance structures where critical agent actions require oversight or approval.

Imagine a legal document summarization agent. It must avoid introducing biased interpretations or removing critical disclaimers that could mislead users. Ensuring neutrality in tone and content is crucial to maintain trustworthiness.

Best Practices

- Annotate datasets with ethical labels (e.g., sensitive content, bias flags).

- Use techniques like counterfactual data augmentation to balance training data.

- Regularly audit your agents using fairness evaluation metrics.

Navigating Regulatory Compliance

Building applications that comply with global standards (GDPR, HIPAA, AI Act) is crucial for sustainable deployment.

Steps to Ensure Compliance

- **Data Privacy:** Anonymize sensitive data before feeding it into the chain. Use pseudonymization or hashing where needed.

- **Consent Management:** Store user consent records if personal data is used. Allow users to revoke consent easily.

- **Audit Trails:** Log agent actions and intermediate outputs for auditing. Maintain version control for prompts and chain configurations.

A healthcare chatbot built with LangChain must ensure that patient information is redacted before being processed by external models. Compliance with HIPAA also mandates encryption of all communications and maintaining secure storage.

Checklist

- Data encryption at rest and transit

- Clear privacy policy disclosures

- Mechanisms for data deletion upon request

… CHAPTER 11   BEST PRACTICES AND PRACTICAL ASPECTS

Optimizing and Scaling LangChain Applications

Performance Tuning Strategies

For production-grade LangChain applications, responsiveness and reliability are vital.

- **Prompt Optimization:** Use concise prompts with clear task instructions. Avoid ambiguous language that could trigger unnecessary agent steps.

- **Caching:** Implement retrieval and output caching to avoid repeated computations. Consider pre-caching results for high-traffic endpoints.

- **Early Termination:** Design agents to recognize when further action is unnecessary. Save compute cycles by embedding confidence thresholds.

Advanced Techniques

- Model distillation to create smaller, faster variants of LLMs for sub-tasks.

- Chain pruning to eliminate unnecessary intermediate steps.

A knowledge base assistant can store commonly asked queries in a cache, enabling direct responses without triggering a complete retrieval pipeline for each request. Smart cache invalidation, driven by content updates, maintains response freshness while preserving performance efficiency.

Scaling Solutions for Production Readiness

Scaling LangChain applications requires addressing computational and architectural challenges proactively.

Techniques for Scalability

- **Distributed Agents:** Deploy agents across multiple instances. Use load balancers to manage distribution.

- **Asynchronous Processing:** Use async chains for I/O-heavy tasks like API calls, database queries.

- **Horizontal Scaling:** Integrate with container orchestration tools like Kubernetes. Use auto-scaling rules based on request volumes.

- **Model Parallelism:** Split large models across GPUs when necessary.

An e-commerce recommendation agent can be horizontally scaled to handle Black Friday traffic spikes without affecting response times. The use of serverless compute for lightweight agents handling minor tasks can further optimize costs.

Tip Monitor latency and throughput continuously. Also, implement circuit breakers to handle service degradation gracefully.

Avoiding Common Pitfalls

Typical Mistakes in LangChain Projects

Common Issues

- **Over-Engineering:** Adding unnecessary agent complexity can introduce new points of failure without meaningful benefit.

- **Prompt Leakage:** Exposing sensitive internal logic or credentials inadvertently through prompts.

- **Evaluation Neglect:** Failing to regularly measure chain quality can lead to silent performance decay over time.

A travel planner agent became confusing because it spawned three sub-agents for simple tasks like suggesting nearby hotels. Users preferred a single, straightforward interaction.

Other pitfalls include failure to distinguish between retrieval-augmented generation (RAG) and basic QA tasks, leading to bloated, confusing chains.

Proactive Strategies for Risk Mitigation

- **Regular Evaluations:** Use evaluation datasets to test chain performance periodically. Leverage LangSmith for tracking metrics like factual accuracy, task success rates.

- **Monitoring:** Implement tracing to diagnose failures in complex chains. Capture inputs, intermediate steps, outputs, and latencies.

- **Fallback Plans:** Always design graceful degradation paths if agents fail. Use fallback intents to route to human agents if automated resolution isn't possible.

Tip Use anomaly detection models to flag unusual agent behaviors proactively before they escalate into larger issues.

The Future of LangChain and AI Agents
Emerging Technologies Shaping LangChain

Several emerging trends are poised to transform LangChain-powered applications:

- **Multi-agent Collaboration:** Teams of agents negotiating or sharing tasks dynamically based on context

- **Autonomous Agents:** Self-reflective agents capable of setting sub-goals, learning from failures

- **Integration with Decentralized Systems:** Decentralized storage (e.g., IPFS) and execution (e.g., blockchain smart contracts) for greater resilience and transparency

A multi-agent research system where one agent gathers sources, other critiques them for credibility, and a third synthesizes a balanced summary. These agents can dynamically reassess task priorities based on emerging information.

New Frontiers

- Emotionally aware agents that adapt tone based on user sentiment.
- Self-improving workflows through reinforcement learning and active fine-tuning.

LangChain Roadmap and Community Vision

The LangChain community aims to

- Enhance **agent memory** architectures for long-term contextual understanding.
- Provide **native support** for emerging LLMs beyond OpenAI and Anthropic.
- Expand **LangSmith** evaluation capabilities for real-world, production-grade pipelines.
- Foster **domain-specific templates** to accelerate industry adoption (e.g., healthcare, finance, legal).

Community Involvement

- Regular contributor meetups and hackathons
- Open bounty programs to develop key missing components

Note "LangChain envisions a future where intelligent agents are accessible, transparent, and adaptable to a wide range of human needs."

Preparing for the Next Generation of AI Development

Embracing Continuous Learning

Staying updated is critical as AI evolves rapidly. Key areas for LangChain developers:

- **Retrieval-Augmented Generation (RAG) Techniques:** Deepen understanding of retrieval, chunking, and hybrid retrieval strategies.

- **Evaluation Metrics:** Move beyond basic accuracy to understand metrics like truthfulness, consistency, and usability.

- **Tooling:** Stay proficient with new integrations such as LangGraph for multi-agent orchestration and LangServe for deployment.

Subscribing to LangChain weekly updates, participating in open source contributions, and engaging in LangChain challenges can significantly accelerate learning.

Checklist

- Follow LangChain GitHub repo (https://github.com/langchain-ai) for updates.

- Join LangChain Slack or Discord communities.

- Attend AI/ML conferences focusing on agent development.

Staying Competitive in a Rapidly Evolving Landscape

Strategies

- **Diversify Skills:** Expand into areas like vector databases (e.g., Qdrant, Weaviate), graph databases (e.g., Neo4j), real-time event streaming (e.g., Kafka).

- **Build Community Presence:** Share LangChain learnings via blogs, talks, or open source projects. Visibility enhances credibility.

- **Stay Agile:** Be ready to pivot strategies and architectures based on evolving best practices.

- **LangChain Documentation**: Refer to original Langchain documentation (https://python.langchain.com/docs/introduction/) for new updates and clarifications.

CHAPTER 11 BEST PRACTICES AND PRACTICAL ASPECTS

Key Takeaways

- Ethical AI requires fairness, transparency, and compliance.
- Scalability involves thoughtful design, monitoring, and testing.
- Pitfalls such as prompt drift, memory overload, and model hallucination can be avoided.
- The LangChain ecosystem is evolving rapidly, with new tools and capabilities.
- Staying updated and iterative is crucial for long-term success.

Index

A

Access controls, 168
Agent architectures
    multi-agent architectures, 207
    plan-and-execute
        architecture, 207
    ReAct architecture, 206
    reflection-based agents, 207, 208
    reflexion architecture, 206
Agents, AI, 90, 91
    callbacks and logging, 102, 103
    customization options, 100
    execution and decision-making
        process, 98–101
    extending capabilities, 100
    types, 91–97
    use cases, 91
Agent state, 204–207
ANN, *see* Approximate Nearest
    Neighbor (ANN)
API-based models, 81
Approximate Nearest Neighbor
    (ANN), 157
ArticleGeneratorChain, 31
Artificial intelligence
    ethical considerations, 229, 230
    regulatory compliance, 230
Azure key vault, 221
Azure OpenAI, 14
Azure OpenAI language
    model, 15
AzureOpenAI model, 13, 14

B

Batch loading, 141
Batch processing, 153
Bias mitigation, 169

C

Callbacks, 102, 103
Callback system, 61
chain.invoke() method, 7, 15, 24, 25
Chains, 12, 13
CharacterTextSplitter, 142
Chatbots, 66, 176–178
    chain, 122
    context awareness, 125, 126
    context maintenance, 121–125
    context management, 115, 116
    conversation flows, 119, 120
    customization, 123–125
    interface, 122, 123
    LangChain environment, 121
    language model, 121
    modular architecture, 117
    multiple LLMs, 118
    purpose, 121
    scalability, 118
    tool integration, 116, 117
Chat models, 61
    configuring and fine-tuning, 105–108
    and large language models (LLMs),
        102, 103
    supported chat models, 103, 104

INDEX

ChatPromptTemplate, 209
Chunking strategies
    fixed-length chunking, 144, 145
    generator's output, 144
    optimization, 147–149
    overlapping, 147, 148
    semantic chunking, 145, 146
    sentence-and paragraph-based, 146
Chunk size, 147–149
CI/CD pipelines, 221
Cloud data, 135
CMS, *see* Content Management Systems (CMS)
Code generation, 5
Code snippet, 20
Cohere embeddings, 151, 152
Compliance, 170, 171
Conditional logic, 13
Connector, 21
Consistency, 71
Constraints, 165
Content filtering, 168
Content Management Systems (CMS), 135
Context awareness, 125, 126
    Chatbots, 126
    complex queries, 126–128
    LangChain, 125
    travel assistant query, 127, 128
Context preservation, 147
Continuous learning, 235
Conversational agent, 93, 94
ConversationBufferMemory, 74–76
ConversationChain module, 122
Conversation flows, 120
    best practices, 120
    components, 119
ConversationSummaryMemory, 76–78
Custom Agents, 95–97

Custom deployments, 175, 176
Custom formatting, 65
Custom output parsers, 67–69
Custom parser, 67–69

D

Data loading, 141
Data privacy regulations, 170, 171
Data validation, 222
Debugging
    assistants, 5
    and monitoring, 35
    techniques, 32
    and testing chains, 36, 37
    and traceability, 124
Dense retrieval methods, 157
Distributed computing, 153, 154
Dynamic prompt generation, 39, 40

E

e-commerce, 232
Embedding data
    chunks, 152
    cohere, 151, 152
    example workflow, 152
    hugging face models, 151
    LangChain, 149–154
    OpenAI, 150
    sentence transformers, 150
Embeddings
    AI applications, 80
    custom models, 83, 84
    in LangChain, 81
    OpenAI, 84
    pre-trained API models, 82, 83
    proprietary, 83, 84
    S-source, 82, 83

Embedding vector, 152
Ensemble methods, 162, 163
Environment isolation, 221, 222
Error handling, 35, 36, 69–71, 110, 111, 119, 226

F

Facebook AI Similarity Search (FAISS), 155
FAISS, *see* Facebook AI Similarity Search (FAISS)
Fine-tuning retrieval models, 161, 162
Fine-tuning technique, 191–198
Fixed-length chunking, 144, 145
Function calling, 53–60

G

Generative Pre-trained Transformer (GPT), 2
GoogleSearchRun, 44
GPT, *see* Generative Pre-trained Transformer (GPT)
Granular queries, 148
Graph-based AI workflow, 178

H

Hierarchical Navigable Small World (HNSW), 157
HNSW, *see* Hierarchical Navigable Small World (HNSW)
HuggingFace, 82
Hugging face models, 151
Human-in-the-loop agents, 4
Hybrid approaches, 189
Hyperparameters, 105

I

Indexing
   scalability and performance, 155
   vector indices, 155
   vector stores, 154
Inference, 197
Integrate parsers, 71–73
Integrating language models, 15, 16
Intent recognition, 119

J

JsonListKeysTool, 44

K

Knowledge source, 134, 135

L

LangChain, 64, 67, 154, 158
   agents, 91–97
   AI workflows, 88
   API integration, 211, 212
   applications, 231
   architecture, 222, 223
   artificial intelligence development, 1
   AzureOpenAI model, 13, 14
   chains and models, 71–73
   chatbots (*see* Chatbots)
   common issues, 232
   components, 12
   custom tool, 212, 213
   development, 5
   ecosystem, 7–9, 173, 174, 209
   embedding models, 81–84
   external APIs, 211
   features and capabilities, 2–4

INDEX

LangChain (cont.)
  fine-tuning, 192
  hybrid architectures, 216, 217
  installation and setup, 5, 6
  integration with, 176
  language models, 2
  memory, 73–80
  memory modules, 115
  ML models, 188–191
  and natural language processing (NLP), 183
  optimization for production, 224, 225
  philosophy, 1
  power, 6, 7
  powered applications, 233
  preparation, 221, 222
  proactive strategies, 233
  for production, 221, 222
  PyTorch, 214–216
  router example code, 217–220
  scaling, 231, 232
  sentiment analysis, 183–191, 214–218
  TensorFlow, 214–216
  testing and evaluation, 225, 226
  tool in agent, 213, 214
  tools and function
    built-in, 44, 45
    custom tool creation, 46–50
    function calling, 53–60
    integrating external API, 49–53
    purposes, 43, 44
  types
    chain design, 32, 33
    custom chains, 29–31
    debugging, 36, 37
    divide-and-conquer approach, 28
    dynamic prompt generation, 39, 40
    integrating language models, 15, 16
    MapReduceChain, 25–29
    multi-prompt routing, 20–22
    performance optimization, 33–35
    prompt libraries, 42–44
    prompt optimization techniques, 40, 41
    robust error handling, 35, 36
    RouterChain, 18–20
    semantic similarity, 23–25
    sequence of operations, 21, 22
    sequential chain, 16–18
    sequential processing, 14, 15
    testing chains, 36, 37
    transformChain, 24, 25
  use cases, 4, 5
  vector stores, 84–89
LangChain Expression Language (LCEL), 61, 107
  advanced chain, 112–114
  error handling, 110, 111
  parallel operations, 109, 110
  with retriever, 111, 112
  sophisticated chains, 108
  syntax, 107
LangChainroadmap, 234
LangGraph, 173, 174, 178
  Agent, 202, 203
  agentic workflows, 179
  Agent state, 204–207
  AI agent, 199
  automated customer support, 180
  autonomous agents, 203
  autonomy and reasoning, 200
  decision support systems, 180
  edges and conditional routing, 201, 202
  graph-based execution, 178
  graphs, 200

LangChain deployments, 178, 179
multi-step reasoning, 199
nodes, 201
SQL queries, 208
state management, 200
text-to-SQL generator, 208, 209
LangServe, 173, 174
advantages, 178
chatbots, 176–178
custom, 176, 177
LangSmith, 173, 174, 180
AI development, 181
with LangChain deployments, 180, 181
project management, 181, 182
Large language models (LLMs), 1, 11, 64, 91, 115, 131
chat models, 102, 103
coherence and relevance, 167, 168
components, 12
multi-turn conversations, 165, 166
prompt engineering, 164, 165
prompts, 11, 37
retrieved context, 164
retrieved information and model knowledge, 166, 167
sensitive information, 168, 169
single API, 11
transparency and explainability, 170, 171 *See also* Agents, AI
Large-scale embedding tasks
batch processing, 153
distributed computing, 153
persistent storage, 153
LCEL, *see* LangChain Expression Language (LCEL)
LLMRouterChain, 21
LLMs, *see* Large language models (LLMs)
Logging, 102, 103

Long-term memory, 77–80

M

MapReduceChain, 25–29
Memory in LangChain
and agents, 74, 76, 77
AI systems, 73
components, 61
inputs and outputs, 74–76
long-term memory, 77–80
types, 73, 74
Memory systems, 3, 8
Metadata-based re-ranking, 160
Metadata filtering, 159
Microservices deployment, 223
Model retrieval improvement
feedback mechanisms, 161
fine-tuning, 161, 162
query expansion and reformulation, 159–163
re-ranking, 160
Model selection strategies, 189–192
Modular component architecture, 3
Monolithic architecture, 222, 223
Monolithic language models, 4
Multi-class classifier, 185
Multi-prompt routing, 20–22
Multi-turn dialogues, 128–130
MyCustomParser, 71

N

Natural language processing (NLP), 2, 37, 119, 183
fine-tuning, 191–198
sentiment analysis, 183–191
Nodes, 201

INDEX

O

Open-source models, 81
OpenWeatherMap API, 212
Output parsers, 61, 106
    appropriate, 64, 65
    chat models, 62
    custom output parsers, 67–69
    error handling, 69–71
    integrating parsers, 71–73
    practical applications, 65–67
    structured output parsing, 64, 65
    types, 62, 63

P

Parallel execution, 13
Parallelization, 141
Performance tuning strategies, 231
Persistent storage, 153
Pipeline orchestrator, 135, 136
Placeholder, 21
Prompt management, 3, 42
Prompt templates, 11, 15, 22
    in large language models (LLMs), 37
    pre-designed structures, 38–40
PydanticOutputParser, 65
Pydantic parser, 63
PyPDFLoader module, 143
PythonAstREPLTool, 44
PythonREPL, 44
PyTorch model, 214, 215

Q

QuerySQLDataBaseTool, 45

R

RAG, *see* Retrieval-augmented generation (RAG)
ReAct framework, 91, 92
Real-time data retrieval, 3, 43
Real-time monitoring, 226, 227
Rephrasing, 160
Re-ranking, 160
Retrieval-augmented generation (RAG), 3, 130, 155, 235
    components, 133
        document processing pipeline, 136
        generation pipeline, 137
        generator, 133, 134
        pipeline orchestrator, 135, 136
        query processing pipeline, 136
        retriever, 133
        storage and retrieval, 136
    data loading, 140–142
    implementation, 137, 138
    multifaceted, 131, 132
    preprocessing, 140–142
    use cases and applications, 138–140
Retrieval efficiency, 147
Retrieval system, 8
Retrieval techniques
    custom retrieval methods, 158
    hybrid retrieval, 158, 159
    metadata filtering, 159
    model retrieval improvement, 159
    similarity search algorithms, 156, 157
Right vector store, 155, 156
Robustness in chains, 35, 36
RouterChain, 18–20
Router example code, 217–220
Routing logic, 205

S

Scaling applications, 223, 224
Schema Definition, 65
Semantic chunking
    techniques, 145, 146
Semantic search, 89, 90
Semantic similarity, 23–25
Sentence transformers, 150
Sentiment analysis, 183–191, 214, 215
Sequential processing, 13–15
SerpAPI, 44
Similarity matching, 89, 90
Sparse retrieval methods, 157
Specialized models, 83
Str Parser, 62
StructuredOutputParser, 65
Structured output parsing, 187
Systematic model evaluation, 191
System prompts, 105

T

Templates, 7
TensorFlow, 83
Text cleaning, 141
TextLoader module, 142
TextRequestsWrapper, 45
Text-to-SQL Generator, 208, 209
Tokenization, 142
Token-limited memory, 78–80
Tools-enabled agent, 94, 95
TransformChain, 24, 25
Transparency, 170, 171
Travel assistant query, 127, 128
TypedDict, 208

U

Uniformity, 145

V

VectorStoreQATool, 45
VectorStoreQAWithSourcesTool, 45
Vector stores
    description, 85
    in LangChain, 85–89
    semantic similarity, 84
Vertical scaling, 224

W

WeatherTool, 214
WebBaseLoader module, 143
WikipediaQueryRun, 44

X, Y

XML parser, 62

Z

Zero-shot ReAct Agent, 91, 92

GPSR Compliance

The European Union's (EU) General Product Safety Regulation (GPSR) is a set of rules that requires consumer products to be safe and our obligations to ensure this.

If you have any concerns about our products, you can contact us on

ProductSafety@springernature.com

In case Publisher is established outside the EU, the EU authorized representative is:

Springer Nature Customer Service Center GmbH
Europaplatz 3
69115 Heidelberg, Germany